Business, Government and the SDGs

Mahmoud Mohieldin · Sameh Wahba ·
Maria Alejandra Gonzalez-Perez ·
Miral Shehata

Business, Government and the SDGs

The Role of Public-Private Engagement in Building a Sustainable Future

palgrave
macmillan

Mahmoud Mohieldin
Cairo University
Giza, Egypt

Maria Alejandra Gonzalez-Perez
Universidad EAFIT
Medellin, Antioquia, Colombia

Sameh Wahba
World Bank
Washington, DC, USA

Miral Shehata
Independent Economic Researcher
Giza, Egypt

ISBN 978-3-031-11195-2 ISBN 978-3-031-11196-9 (eBook)
https://doi.org/10.1007/978-3-031-11196-9

This Palgrave Macmillan imprint is published by the registered company Springer Nature Switzerland AG
The registered company address is: Gewerbestrasse 11, 6330 Cham, Switzerland

Contents

About the Authors

Mahmoud Mohieldin[1] Professor of Economics, Cairo University; the UN Special Envoy on Financing the 2030 Agenda for Sustainable Development; the UN 7th High-Level Champion for Climate Action; Former Senior Vice President of the World Bank Group; and Former Minister of Investment in Egypt.

Sameh Wahba[2] Regional Director for Sustainable Development, Europe and Central Asia Region, The World Bank and former Global Director, Urban, Disaster Risk Management, Resilience and Land Global Practice.

Maria Alejandra Gonzalez-Perez Full Professor of Management and Sustainability at Universidad EAFIT, Colombia; and Member of the SDG 1 (no poverty) Global Council.

[1] The views expressed herein are entirely the author's and should not be attributed to the United Nations or the International Monetary Fund, its Executive Board or management.

[2] The findings, interpretations, and conclusions expressed in this book are entirely those of the authors. They do not necessarily represent the views of the International Bank for Reconstruction and Development/The World Bank and its affiliated organizations, or those of the Executive Directors of the World Bank or the governments they represent.

Miral Shehata[3] Independent Economic Researcher.

[3] The views, findings, and interpretations expressed in the book should be entirely attributed to the author.

List of Figures

List of Tables

CHAPTER 1

SDGs and the 2030 Agenda: On Crisis and Opportunity

Abstract For the first time in history, the entire human population encountered a perfect storm. This consisted of a health emergency caused by the COVID-19 pandemic; economic contraction accompanied by demand and supply shocks as a result of lockdowns; massive job and livelihood losses; disruptions to global trade and supply chains; severe financial and commodity market volatility; and a compounding environmental crisis caused by climate change. Many Sustainable Development Goals (SDGs) will not be met by 2030 if such challenges are not addressed. This challenge involves reconsidering the roles of government, businesses, and cities in accomplishing the SDGs and the 2030 Agenda, as well as their conventional partnerships and domains of influence. Global challenges aren't limited to pandemics and endemics. Climate change, biodiversity loss, food scarcity, armed conflicts, climate-induced migration, and political crises all pose threats to the SDGs. However, these crises are exactly the vulnerabilities and disruptions that the SDGs seek to tackle. The SDGs provide a roadmap to enhance system preparedness and invest in resilience to strengthen systems ability to weather shocks and bounce back. In this sense, it is important to analyze the roles assumed by national governments, businesses, and local and regional governments (LRGs) in advancing the 2030 Agenda, as well as the interactions among these actors to ensure their coordinated and coherent responses to crises and sustainable investment in the necessary systems.

M. Mohieldin et al., *Business, Government and the SDGs*,
https://doi.org/10.1007/978-3-031-11196-9_1

Keywords SDGs · COVID-19 · Crisis recovery · Partnerships · Governance · Resilience · Building Forward Better

Introduction: COVID-19 and Sustainable Development

The year 2020 marked a turning point for the entire human population. The health emergency caused by the spread of the COVID-19 pandemic spiraled; there was economic contraction accompanied by demand and supply shocks as a result of lockdowns, disruptions to global trade and supply chains, severe financial and commodity market volatility, massive job and livelihood losses, and significant disruptions to learning gains. The pandemic halted or even reversed years of development progress, with 2020 witnessing a rise in global extreme poverty for the first time in over 20 years. Inequalities within and among countries were also further intensified; women disproportionately suffered a larger share of job losses, and the livelihoods of 1.6 billion workers in the informal economy were threatened, thus highlighting how the pandemic imposed a differentiated impact on the poorest and most vulnerable groups (United Nations, 2021). These effects were emphasized given the persistence of the climate crisis, as the world has remained off track in meeting the pledges of the Paris Agreement, and how there remains a continuous loss of biodiversity and degradation of terrestrial ecosystems at alarming rates (United Nations, 2021).

The pandemic poses a major setback to achieving the 2030 Agenda. In 2020, the global average SDG Index, which measures the overall global performance of the 17 SDGs, witnessed the first decline since the adoption of the SDGs in 2015. With the numbers for the year 2021 already out, it appears that 2021 was the second consecutive year the world was no longer making progress on the SDGs, as the average SDG Index had slightly score decreased. This decline is partially driven by slow or non-existent recovery in poor and vulnerable countries and by the many overlapping health, economic, and security crises. For instance, performance on SDG 1 of No Poverty and SDG 8 of Decent Work and Economic Growth are still below pre-pandemic levels, especially in many low-income and lower-middle-income countries. These major setbacks are particularly worrisome as even prior to the pandemic the world was making slow progress toward SDGs, at a rate that was then unlikely to reach the targets set for 2030 (Sachs et al., 2021, 2022).

COVID-19 exposed the root causes of vulnerabilities and unpreparedness in our health and economic systems. It reflected the deep-rooted problems in our societies, including insufficient social protection, unprepared health systems, structural inequalities, unsustainable patterns of production and consumption, environmental degradation, and climate change. The pandemic also demonstrated the interlinkages and interdependency among the dimensions of sustainability—ranging from health and social and economic prosperity to climate issues and ecosystems (United Nations, 2021). This, in turn, emphasizes the urgency to follow a coherent multidimensional, and systematic approach rather than a sector-by-sector sequential approach in tackling crises in general (Mohieldin & Shehata, 2021). By its original multidimensional and integrated design, the 2030 Agenda can act as a roadmap to crisis recovery and building resilience to future shocks, which is even highlighted by recent global developments such as the Russian-Ukrainian crisis. The 2030 Agenda and its 17 SDGs allow countries to focus on the needed factors of inclusion, equity, growth, resilience, and sustainability, all the while recognizing the risks and complex links between social, environmental, and economic dimensions.

Focusing on the SDGs as a roadmap for recovery and resilience requires capitalizing on the lessons the world has learned from past crises, the pursuit of the Millennium Development Goals (MDGs), and the past six years of efforts toward the 2030 Agenda. This entails revising the responsibilities of the different actors in the sustainable development policy sphere and analyzing their interactions to understand the potential to advance the 2030 Agenda as well as highlight the gaps in their assumed roles. As such, the focus of this text will center on this in relation to national governments, local and regional governments (LRGs), and the private sector.

From the MDGs to the SDGs

At a high-level meeting of the United Nations General Assembly (UNGA) in 2010 to review the implementation of the Millennium Development Goals (MDGs)—the first United Nations time-bound development goals—, there was a consensus that governments needed to plan ahead to advance sustainable development in the post-2015 era. This was followed by the Rio+20 Conference in June 2012, which had the primary objective to address emerging challenges and secure political commitment for sustainable development while assessing both the progress to date and the major gaps in the implementation of outcomes from major summits on sustainable development. A major point in the Rio+20 outcome document, *The Future We Want*, called for the establishment of an 'inclusive and transparent intergovernmental process on sustainable development goals that is open to all stakeholders, with the motivation of developing global sustainable development goals that will be agreed upon by the General Assembly' (United Nations, 2012). This resulted in the establishment of the Open Working Group (OWG), which held thirteen sessions from March 2013 to July 2014 to develop the SDGs and targets. This process benefitted from the input of academics, civil society organizations (CSOs), think tanks, and the private sector.

Following the work of the OWG, the UNGA formed intergovernmental negotiating processes to agree on the outcome document for the 2015 United Nations Sustainable Development Summit. This was necessary to declare the SDGs, clarify their means of implementation, and establishment of a follow-up and review process. The negotiations, which were concluded in July 2015, had to consider two parallel processes: the climate change negotiations under the UN Framework Convention on Climate Change (UNFCCC) and the Third International Conference on Financing for Development, whose outcome document, the Addis Ababa Action Agenda (AAAA) serves as the means of implementation for the 2030 Agenda. At that point, it was clear that the magnitude of financial resources, technological transfer, and capacity building needed to achieve the 2030 Agenda would be unprecedented. Accordingly, the AAAA comprises a global framework for financing development by harnessing and coordinating all domestic, external, public, and private financial flows while stressing the role of science, technology, innovation, capacity building, and data and monitoring (Bettelli, January 2021; Chasek et al., 2016; United Nations, 2015a).

Besides the participatory, multistakeholder process underlying the formulation of the 2030 Agenda, the world had extracted some lessons on how to accelerate developmental progress through its engagement with the MDGs. These lessons were further clarified via the series of MDG acceleration reviews conducted by the United Nations System Chief Executives Board for Coordination (CEB) during the final years of the MDGs. The purpose of these acceleration reviews was to address the problem of the underachievement of the MDGs by discussing bottlenecks hindering progress, methods of tackling such bottlenecks, mechanisms to scale-up successful policies, and initiatives to accelerate progress. The reviews had a number of emerging recurrent themes: the need to tackle inequality, invest in boosting resilience, promote human rights and women's empowerment, enhance focus on displacement-induced migration and fragility issues, and to facilitate the transition from humanitarian assistance to development. Additionally, the takeaways from these reviews highlighted what is needed for progress toward development goals. One of which is how development goals are interrelated, and thus call for the adoption of comprehensive, multisectoral policies. These involve delivery at the subnational level to achieve progress, the efficient allocation of resources supported by good governance and strong institutions to maximize the return, and the availability of quality data for monitoring purposes, policy development, and designing targeted interventions (Martínez-Solimán & Mohieldin, 2016).

The first lesson learned from the MDGs is that even when development goals are expressed as sector-specific, they cannot be achieved in sectoral silos. Also, the policy landscape surrounding them is usually characterized by cross-sectoral synergies and trade-offs. To elaborate, well-designed social protection programs can support households accessing better health, education, and nutrition. Additionally, there are multiple understandings of what integration means and how it can be capitalized on. In some instances, integration implies that progress on one goal or target can have a positive impact on another, or that interventions in certain domains can have multiplying effects on other sectors. The second lesson gained from the MDGs would be that effective policy design and implementation hinges on strong government involvement as well as high political buy-in and commitment to the development goals. Such government ownership has the merit of ensuring that efforts to realize the desired outcomes have the support of the institutions and the required means of implementation. Furthermore, strong institutions and adequate policies are proven to be key for strengthening the quality of public spending directed toward accelerating the progress toward development goals. This highlights the third lesson: high-impact policies and programs are the ones for which resources are allocated, and the inadequate prioritization of development goals in the budgeting process can create a continued deficit in implementation (Martínez-Solimán & Mohieldin, 2016).

The last lesson from the reviews demonstrates that achieving results in development necessitates covering all targeted populations and ensuring last-mile service delivery. Also, progress acceleration requires interventions at the subnational level. For example, the degree of progress toward the MDGs varied across rural and urban regions; people living in cities witnessed far more development progress than those residing in rural areas, owing to the role of scale economies in urban centers and the bottlenecks of service provision in sparsely populated rural regions. On a similar note, the reviews showcased that focusing on the vulnerable populations, especially in fragile settings, calls for bridging the humanitarian and development agendas within one coherent policy of engagement, and the adoption of a holistic perspective of development as a continuum. This would include both addressing the immediate needs of the populace and long-term development measures to build resilience. Reaching the most vulnerable groups and designing evidence-based policies also require timely, disaggregated, and high-quality data. Significant data gaps were among the most acute challenges confronting developing countries in the context of the MDGs, particularly because data generation necessitates investing in capacity building and infrastructure, which has not always been an option for these states. Consequently, besides traditional data sources, there has been a continuous need to explore new technology and open data to revolutionize the capacities of countries for data collection. This would involve interactions between the public and private sectors as well as innovations regarding leveraging technology for development data (Martínez-Solimán & Mohieldin, 2016). These CEB reviews thus exhibited that lessons from the MDGs which should feed into the post-2015 Agenda can be summarized as follows: the presence of comprehensive policy frameworks, effective implementation including last-mile delivery, adequate and efficient allocation of finance and resources supported by strong institutions, and dependable data that allows for monitoring and feeding into designing evidence-based policies.

The Adoption of the SDGs and the Leading Role of National Governments

At the onset of the United Nations Sustainable Development Summit convened in September 2015, all the 193 Member States of the United Nations unanimously adopted the *Transforming Our World: The 2030 Agenda for Sustainable Development* (United Nations, 2015b). The 2030

Agenda emphasizes that each state is responsible for its social and economic development. Accordingly, it tasks governments with setting their own national targets guided by the global goals of the 2030 Agenda by taking into account their respective national circumstances. In this task, each government then has to decide the method by which to incorporate these global targets in its national planning processes, strategies, and policies, as well as to realize the link between sustainable development and other ongoing processes in the economic, social, and environmental fields. Governments are thus critical actors to enable policy coherence and strengthen links across sectors in light of the integrated and comprehensive design of the 2030 Agenda. It is thus maintained that governments have the primary responsibility of implementing the SDGs and ensuring follow-up and review over the 15 years-period spanned by the 2030 Agenda (Allen et al., 2018; Dalby et al., 2019; United Nations, 2015b).

Six years after the adoption of the 2030 Agenda, complex institutionalization of the SDGs in national settings can be seen, as the former achieved relatively high visibility as an overarching policy agenda in both developing and developed countries. The majority of countries have developed strategies and action plans to implement the SDGs. However, there is no uniform mechanism of doing so; some countries developed a national sustainability strategy that is directly linked to the 2030 Agenda while other governments followed a mainstreaming approach and have the relevant SDGs implemented by the appropriate government ministry. Still, major differences are evident across countries in terms of the depth of SDG institutionalization (Sachs et al., 2022; UN DESA, 2021). Additionally, there appears to be some progress in the institutionalization of SDG follow-up and review systems, as well as in identifying sets of nationally relevant indicators. Many countries have identified the institutions responsible for SDG monitoring, but the performance of this institutionalization has not always resulted in an effective follow-up review. Nevertheless, countries have made progress in the preparation of the Voluntary National Reviews (VNRs), which can then act as a proxy to assess their commitment to the SDGs (Table 1.1).

Effective public institutions and good governance are also important in guaranteeing the highest return and lowest efficiency losses in public infrastructure investments, which is indispensable for the realization of the SDGs. For instance, poor infrastructure governance is estimated to result in efficiency gaps of 15% in advanced economies and as high as 53% in low-income developing countries (Baum et al., 2020).

Table 1.1 Status of the submission of the Voluntary National Reviews

Criteria	*Number of countries*
Countries that have submitted no VNRs	6
Countries that have submitted one VNR	100
Countries that have submitted two VNRs	72
Countries that have submitted three VNRs	13
Countries that have submitted four VNRs	2

Note Numbers include countries conducting a VNR in 2022. The six countries that submitted no VNRs are Haiti, Iran, Myanmar, South Sudan, the United States, and Yemen. The two countries with four VNRs are Togo and Uruguay which committed to be submitting their VNRs in 2022
Source Adapted from Sachs et al., 2022 based on the United Nations, Voluntary National Reviews, Sustainable Development Knowledge Platform

Reliable and effective systems of public administration are also key in ensuring that incremental borrowing and increase in debt stock is channeled to productive investments needed to raise national income rather than toward consumption or wasteful investment (Sachs et al., 2022). Relatedly, governments can impact the SDGs through public finance by providing incentives for more sustainable production and consumption patterns, as well as work to generate revenues that can be used to finance the SDGs. When public budgets are fully aligned with the SDGs, national development policies and plans will be effectively implemented. Otherwise, governments are left with a discrepancy between their expressed political support for the SDGs and unlocking the resources needed to acquire this support. The budget, when combined with the right mix of regulations and macroeconomic stability, acts as a signal for non-state actors to mobilize the necessary investments to fill the SDG investment gap (Sachs et al., 2022; UNCTAD, 2014; UNDP, 2020).

Despite the fact that national governments are the primary actors responsible for the implementation, follow-up, and review of the 2030 Agenda, the ambitions of the goals necessitate that governments work with a diverse set of actors to realize such ambitions. Two such actors that this text focuses on are: local and regional governments (LRGs) and the business sector.

Reaching the Last-Mile and the Role of Local and Regional Governments

The SDGs emphasize the importance of cities and human settlements for implementing the 2030 Agenda's global objectives, highlighting the need for local action that serves global goals and aligns with local profiles (Fenton & Gustafsson, 2017; Valencia et al., 2019). An estimated 60–65% of the SDG targets require LRGs' involvement, two of which—SDG 11 and 16—critically depend on local government institutions (OECD, 2020). Moreover, lessons learnt from the MDGs highlight that it is crucial to consider the spatial nature of development. Focusing only on the thematic goals and failing to see what underlies national averages would result in missing both targeting the last mile, and effective policies and mechanisms that directly tackle exclusionary factors and avoid bottlenecks in service provision (Elmassah & Mohieldin, 2020; Martínez-Solimán & Mohieldin, 2016; UNDP & UNCDF, 2016).

Since LRGs represent the level of government closest to the populace, they usually possess the best understanding of a given community's needs and priorities, and can act to ensure that no one is left behind. Accordingly, LRGs have a central role in service delivery, thus contributing to the SDGs on water, sanitation, housing, transport, education, and health. In many cases, LRGs contribute to economic development through employment opportunities. They are also stepping up to address issues that were previously seen as the responsibility of national governments alone. Importantly, cities are moving to solidify climate action through the creation of the Covenant of Mayors, the C40 cities network, and other initiatives that seek to establish and improve links between local and international governance levels (Mohieldin & Shehata, 2021). Several LRGs have also committed to the follow-up and review of the SDG implementation at the local level, where in 2018 several LRGs became signatories of a declaration committed to carry out Voluntary Local Reviews (VLRs) to gauge how their strategies and programs align with the SDGs.

The well-established role of LRGs in advancing the development agenda has given momentum to localization as a mechanism to align the SDGs with local development priorities. Localization allows for ensuring that decisions are taken in the level of governance closest to the people and thus reflect their needs and goals without compromising the benefits of centralization and capitalizing on capacity, resources, and national security. It is important to ensure that national governments do not perceive localization as a top-down relationship where the SDGs and accompanying national development priorities are shifted down to LRGs in a passive manner. The concept of localization should fulfil the need for vertical integration, which involves consistency, coordination, and collaboration across different levels of government to advance the 2030 Agenda and ensure policy coherence (Elmassah & Mohieldin, 2020; UCLG, 2017).

Environmental, Social, and Governance (ESG) Factors, SDGs, and the Role of the Business Sector

Although it is fundamentally the responsibility of governments to implement the SDGs, it would be impossible to claim that the goals can be achieved without the meaningful contribution of the business sector. Above all, the private sector is a major generator of employment, economic growth, finance, and innovation and technology. Since the business sector operates within and interacts with the surrounding ecosystem, it is reasonable to consider its environmental and social impacts as well as how its governance structure both meets the needs of the different stakeholders and ensures transparency. The business contribution to sustainability is best seen through its adherence to environmental, social, and governance (ESG) factors.

There is a strong business case for the private sector to focus on ESG-related issues as much as it focuses on profit generation. There is also a growing interest in ESG from regulators, as the number of ESG regulations and standards across the globe has almost doubled in the five-year period between 2015 and 2020 (EY & Oxford Analytica, June, 2021). Several governments are redirecting their regulations to encourage ESG investing and mandatory disclosures. For instance, the European Union's legislation on the Corporate Sustainability Reporting Directive (CSRD) will require all large companies to publish regular reports on their social and environmental impact activities. Nearly 50,000 companies fall within the scope of this legislation and they will need to comply with the directive from January 2024. The directive envisions the adoption of EU sustainability reporting standards that are being developed by the European Financial Reporting Advisory Group (EFRAG) (Council of the European Union, 21 June 2022).

Countries have also been announcing mandatory climate risk reporting. For example, New Zealand has recently passed a legislation making climate-related disclosures mandatory for some organizations in line with the recommendations of the Task Force on Climate-related Financial Disclosures (TCFD) (New Zealand Ministry for the Environment, 17 June 2022). Similarly, in January 2022, the United Kingdom passed its Climate-related Financial Disclosure regulations which are broadly in line with the four pillars of the TCFD recommendations (United Kingdom Statutory Instruments, 2022 No. 31). Survey results show that regulatory development is becoming one of the main drivers in embracing ESG concerns, with 91% of respondents to a Barnett Waddingham survey declaring that evolving regulation and legal requirements is a key factor that encourages taking into account ESG considerations in investment decisions (Barnett Waddingham, March 2022). Given this proliferation in regulations and standards coupled with an increased demand for high-quality comparable reporting, the ESG reporting landscape is now quickly conforming to globally harmonized disclosure standards. A major development toward this harmonization is the creation of International Sustainability Standards Board (ISSB) under the International Financial Reporting Standards Foundation (IFRS), with endorsement from the G7, G20 and the International Organization of Securities Commissions (IOSCO). The ISSB is mandated with delivering a comprehensive global baseline of sustainability-related disclosure standards that provide investors and other capital market participants with information about companies' sustainability-related risks and opportunities to help them make informed decisions (Poole & Sullivan, 29 October 2021).

Strong ESG propositions also help companies tap new markets and expand into existing ones. Research has demonstrated that customers are willing to pay for "greener" products (Henisz et al., November, 2019). Results from PwC, 2021 ESG Consumer Intelligence Research show that the majority of consumers believe that companies should be actively adopting ESG best practices. More importantly, around 80% of consumers surveyed claimed they are more likely to buy from companies that share their values with ESG elements, with many consumers indicating they are willing to pay more for healthier, safer, and more environmentally and socially conscious products and brands. Similarly, consumers appear ready to penalize companies that have poor relations with their employees, communities, and environment by discontinuing to buy from them (PwC, 2021). Strong ESG performance is also meeting investors' interests as 89% of institutional investors say that companies that demonstrate strong ESG performance deserve a premium valuation of their share price, and 90% of investors believe that companies that prioritize ESG initiatives represent better opportunities for long-term returns (Edelman, 2020). Moreover, employees are more likely to work for companies which have strong commitments to ESG considerations (PwC, 2021). This is important because according to Deloitte Millennial Survey, by 2025 millennials—the so-called 'values-driven generation'—will represent three-quarters of the global workforce, highlighting that companies that foster innovative thinking and make sustainable positive contributions to society would manage to retain highly skilled employees (Deloitte, 2014). These consumer and employee preferences are shown in the light of the massive availability of information in the digital age. This has enabled employees, suppliers, business partners, members of the community, activists, and society at large, to act as equal participants (stakeholders) in a direct dialogue with companies about their expectations. As a result, generations that have grown up with this availability of information are able to call for accountability and have higher expectations for businesses regarding their commitment to positive change in consideration of the people and planet, as opposed to only profits (Poole & Sullivan, 29 October 2021).

The SDGs, Resilience, and the Response to Crises

The lessons learned from the world's engagement with the MDGs and the responsibilities assumed by different actors in the context of the 2030 Agenda has shed light on the multidimensional nature of environmental, social, and economic issues. Crises like COVID-19 and the more recent war in Ukraine reveal the systematic interlinkages between all the dimensions of sustainable development and expose the vulnerabilities, unpreparedness, and flaws of systems. Consequently, despite how crises have proven to cause major setbacks in countries' commitment to the SDGs, the SDGs themselves provide a roadmap for recovery and fostering resilience against future shocks by their focus on directing investment toward physical and digital infrastructure, human capital accumulation, and resilience. The world needs economic resilience with new fiscal and debt sustainability frameworks, societal resilience with strong social protection schemes and well-prepared service delivery systems, and climate resilience by directing investment toward renewable energy and sustainable infrastructure.

In his book *The Resilient Society*, Economist Markus Brunnermeier discusses the concept of resilience as the ability to weather a storm, recover and have a vision for the new normal. This, resilience can be built by investing in mechanisms that work to acknowledge shocks rather than resist them, which would ultimately help societies to adapt and bounce back. This view of resilience requires an emphasis on the "just in case" approach rather than the "just in time" attitude that currently characterizes our production systems. At its heart, resilience encompasses the need for adaptation. There needs to be acknowledgment of the fact that the world has many disruptions that have diverse consequences. However, the commonality among these disruptions is the need for societies to be prepared and maintain enough flexibility to effectively engage with them (Mohieldin, April 2022).

At the core of crisis response is the concept of "Building Forward Better." Building Forward Better takes into account how there are populations that do not wish to return to their past circumstances or 'build back' what was previously in place, especially if this past is characterized with structural disadvantages and inequalities (UN DESA, n.d.). In Building Forward Better, the aim is to transform humankind's relationship and interaction with the environment, as well as to tackle patterns of social and economic vulnerabilities in order to enhance future resilience.

Acknowledgements The authors appreciate the patience and great support from Marcus Ballanger, Commissioning Editor, Scholarly Business at Palgrave Macmillan. They thank Janina Krieger for her enthusiasm for AI, as well as recognize the input of the AI robot in Chapter three. Additionally, the authors appreciate the efforts of Haajar Abu Ismail, and her excellent editorial contributions to the chapters of this book. The authors would also like to thank their families and beloved ones for their patience and support. This book was written using the time that belonged to them. Furthermore, the authors express their gratitude toward Patricia Holly Purcell, Private Sector Specialist, and to anonymous reviewers for their thoughts regarding the proposal of this book.

References

Allen, C., Metternicht, G., & Wiedmann, T. (2018). Initial progress in implementing the sustainable development goals (SDGs): A review of evidence from countries. *Sustainability Science, 13*, 1453–1467.

Baum, A., Mogues, T., & Verdier, G. (2020). *Getting the most from public investment, in well spent: How strong infrastructure governance can end waste in public investment?* (G. Schwartz, M. Fouad, T. Hansen, & G. Verdier, Eds.). International Monetary Fund.

Barnett Waddingham. (2022, March). *Sustainable investor insights survey.* Available from: https://www.barnett-waddingham.co.uk/comment-insight/research/sustainable-investor-insights-survey/

Bettelli, P. (2021, January). *What the world learned setting development goals.* International Institute for Sustainable Development. Retrieved May 30, 2022, from https://www.iisd.org/articles/deep-dive/what-world-learned-setting-development-goals

Brunnermeier, M. (2021). *The resilient society: Lessons from the pandemic for recovering from the major next shock.* Endeavor Literary Press, ISBN 978-1-7374036-0-9.

Chasek, P. S., Wagner, L. M., Leone, F., Lebada, A.-M., & Risse, N. (2016). Getting to 2030: Negotiating the post-2015 sustainable development agenda. *RECIEL, 25*, 5–14. https://doi.org/10.1111/reel.12149

Council of the European Union. (2022, June 21). *New rules on corporate sustainability reporting: Provisional political agreement between the Council and the European Parliament.* Press Release. Available from: https://www.consilium.europa.eu/en/press/press-releases/2022/06/21/new-rules-on-sustainability-disclosure-provisional-agreement-between-council-and-european-parliament/

Dalby, S., Horton, S., & Mahon, R. (2019). Global governance challenges in achieving the sustainable development goals. In S. Dalby, S. Horton, R. Mahon & D. Thomaz (Eds.), *Achieving the sustainable development goals: Global governance challenges* (1st ed., pp. 1–18). Routledge Studies in Sustainable Development. Routledge

Deloitte. (2014). *Big demands and high expectations: The Deloitte Millennial Survey.* Executive Summary. Available from: https://www2.deloitte.com/content/dam/Deloitte/global/Documents/About-Deloitte/gx-dttl-2014-millennial-survey-report.pdf

Edelman. (2020). *Edelman Trust Barometer 2020 Special Report: Institutional Investors, U.S. Results.* Available from: https://www.edelman.com/sites/g/files/aatuss191/files/2020-11/Edelman%202020%20Institutional%20Investor%20Trust_FINAL.pdf

Elmassah, S., & Mohieldin, M. (2020). Digital transformation and localizing the sustainable development goals (SDGs). *Ecological Economics, 169*, 106490. https://doi.org/10.1016/j.ecolecon.2019.106490

Ernst & Young (EY) & Oxford Analytica. (2021, June). *The future of sustainability reporting standards: The policy evolution and the actions companies can take today.* Available from: https://assets.ey.com/content/dam/ey-sites/ey-com/en_gl/topics/sustainability/ey-the-future-of-sustainability-reporting-standards-june-2021.pdf

Fenton, P., & Gustafsson, S. (2017). Moving from high-level words to local action—Governance for urban sustainability in municipalities. *Current Opinion in Environmental Sustainability, 26–27*, 129–133.

Henisz, W., Koller, T., & Nuttall, R. (2019, November). Five ways that ESG creates value: Getting your environmental, social, and governance (ESG) proposition right links to higher value creation. Here's why. *McKinsey Quarterly*. Available from: https://www.mckinsey.com/business-functions/strategy-and-corporate-finance/our-insights/five-ways-that-esg-creates-value

Martínez-Solimán, M., & Mohieldin, M. (2016). *Lessons and recommendations from the CEB Reviews: A forward-looking perspective from the MDG era in transitioning from the MDGs to the SDGs*. United Nations Development Programme and the World Bank Group.

Mohieldin, M. (2022, April). Progress, disruption, and the path ahead. *Ahram*. Available from: https://english.ahram.org.eg/NewsContentP/50/464934/AlAhram-Weekly/Progress,-disruption,-and-the-path-ahead.aspx

Mohieldin, M., & Shehata, M. (2021). The SDGs as an operational framework for post COVID-19 response and recovery. *AD-Minister, 38*, 5–42. https://doi.org/10.17230/Ad-minister.38.1

New Zealand Ministry for the Environment. (2022, June 17). *Mandatory climate-related disclosures: The Government has passed legislation making climate-related disclosures mandatory for some organisations. The requirement will apply to large publicly listed companies, insurers, banks, non-bank deposit takers and investment managers.* Available from: https://environment.govt.nz/what-government-is-doing/areas-of-work/climate-change/mandatory-climate-related-financial-disclosures/

OECD. (2020). A territorial approach to the sustainable development goals: Synthesis report. In *OECD urban policy reviews*. OECD Publishing.

Poole, V., & Sullivan, K. (2021, October 29). *Tectonic shifts: How ESG is changing business, moving markets, and driving regulation*. Deloitte Insights. Available from: https://www2.deloitte.com/us/en/insights/topics/strategy/esg-disclosure-regulation.html

PwC. (2021). 2021 ESG Consumer Intelligence Series: *Beyond compliance: Consumers and employees want business to do more on ESG—How business can close the expectations gap*. Available from: https://www.pwc.com/us/en/services/consulting/library/consumer-intelligence-series/consumer-and-employee-esg-expectations.html

Sachs, J., Lafortune, G., Kroll, C., Fuller, G., & Woelm, F. (2022). *Sustainable Development Report 2022: From crisis to sustainable development: The SDGs as roadmap to 2030 and beyond*. Cambridge University Press.

Sachs, J., Schmidt-Traub, G., Kroll, C., Lafortune, G., & Fuller, G. (2021). *Sustainable Development Report 2021: The decade of action for the sustainable development goals*. Cambridge University Press.

United Cities and Local Governments (UCLG). (2017). *National and sub-national governments on the way towards the localization of the SDGs*. Report Towards the 2017 HLPF. Global Task Force of Local and Regional Governments. Available from: https://www.uclg.org/sites/default/files/localgov_report_localizationsdg_hlpf.pdf

United Kingdom Statutory Instruments. (2022, January 17). *The Companies (Strategic Report) (Climate-related Financial Disclosure) Regulations 2022*. Statutory Instrument 2022 No. 31.

United Nations. (2015a). *Addis Ababa action agenda of the third international conference on financing for development (Addis Ababa action agenda)*. United Nations.

United Nations. (2015b). *Transforming our world: The 2030 agenda for sustainable development*. Resolution adopted by the General Assembly on 25 September 2015b. A/RES/70/1. U.N. General Assembly, Seventieth Session. Agenda items 15 and 116. United Nations.

United Nations. (2012). *The future we want*. Resolution adopted by the General Assembly on 27 July 2012. A/RES/66/288. U.N. General Assembly, Sixty-sixth session. Agenda item 19. United Nations.

United Nations. (2021). *The sustainable development goals report 2021*. United Nations.

United Nations Conference on Trade and Development (UNCTAD). (2014). *World investment report 2014: Investing in the SDGs: An action plan*. United Nations.

United Nations Development Programme (UNDP). (2020). *Budgeting for the sustainable development goals: Aligning domestic budgets with the SDGs: A Guidebook*. https://sdgfinance.undp.org/sites/default/files/UNDP%20Budgeting%20for%20the%20SDGs%20-%20Guidebook_Nov%202020.pdf

United Nations Development Programme (UNDP) & United Nations Capital Development Fund (UNCDF). (2016). *Getting to the last mile in least developed countries*. New York.

United Nations Department of Economic and Social Affairs (UN DESA). (n.d.). *It is time to build forward together*. Available from: https://www.un.org/en/desa/it-time-build-forward-together

United Nations Department of Economic and Social Affairs (UN DESA). (2021). *World Public Sector Report 2021: National institutional arrangements for implementation of the sustainable development goals: A five-year stocktaking*. United Nations. Available from: https://www.un.org/en/desa/world-public-sector-report

Valenica, S., Simon, D., Croese, S., Nordqvist, J., Olokoe, M., Sharma, T., Buck, N., & Versace, I. (2019). Adapting the sustainable development goals and the new urban agenda. *International Journal of Urban Sustainable Development, 11–1*, 4–23.

CHAPTER 2

The Leading Role of National Governments in "Building Forward Better"

Abstract The 2030 Agenda emphasizes that national governments have the critical responsibility of implementing the Sustainable Development Goals (SDGs) along with ensuring their follow-up and review. This requires that governments create national and goals that mirror global ambitions and put in place the necessary institutional mechanisms and monitoring systems to translate these goals into effective implementation. Achieving the SDGs requires addressing the questions related to relevant complementarities and trade-offs to the wider policy domains. This in turn requires having in place policy frameworks that foster horizontal integration given that the SDGs and climate action cannot be achieved by sectoral siloed policies. This multisectoral, comprehensive view of sustainability policies needs to be complemented with a whole-of-government approach that engages local and regional governments (LRGs) to ensure last-mile delivery as well as attracts the private sector to the scene. The implementation of policies needs to be backed by strong high-quality disaggregated data that is capable of informing the decision-making process and strengthens delivery within a data governance framework that promotes trust in data transactions. The cross-cutting theme is that realizing the ambitions of the 2030 Agenda hinges on adequate finance, a factor that has been further stressed by the pandemic. Fiscal policies and public financial management are essential for both aligning the public financial system with the SDGs and unlocking the necessary SDG funds from non-state actors, specifically the private sector. The tripartite of

M. Mohieldin et al., *Business, Government and the SDGs*,
https://doi.org/10.1007/978-3-031-11196-9_2

policy framework and effective implementation institutions, dependable data, and adequate finance has been repeatedly stressed in the lessons learnt from the pandemic and other global crises. These are key factors toward investing in infrastructure, resilience, and human capital.

Keywords National governments · COVID-19 · Pandemic · Sustainable Development Goals (SDGs) · Development Policies · Crisis · Building forward better

Introduction

The 2030 Agenda focuses on the essential role of national governments in its implementation. It is fundamentally the state's responsibility to realize these goals and review the progress toward them. Governments are set to work closely with multiple stakeholders including local authorities, subregional institutions, international institutions, academics, the private sector, philanthropic organizations, and volunteer groups as well as others. The 2030 Agenda also entrusts governments with the mission to set their 'own national targets guided by the global level of ambition that takes into account national circumstances' (United Nations, 2015a). Accordingly, governments should have the 2030 Agenda targets incorporated in their national planning processes, policies, and sectoral strategies while recognizing the complexity of synergies and trade-offs in the economic, social, and environmental fields.

The role of national governments is multifold. In their capacity as leaders, their ability to adopt a whole-of-government and multisectoral approach to enhance internal integration and policy coherence will determine their success in dealing with the complex nature of the agenda. Given their duty to lead, governments should act as strategic partners that ensure the engagement of all other stakeholders and mobilize their contributions. This includes designing more resource strategies that incorporate packages that provide financing for development, especially for that sourced from private investors and businesses (Gornitzka & Pipa, April 2018). Achieving progress on the SDGs hence involves broadening the complex equation of complementarities and trade-offs to account for the whole spectrum of policy areas addressed by the SDGs, which in turn requires prioritization. The COVID-19 crisis in combination with both the risk of falling into a deeper cycle of unsustainable debt and the rising intensity of climate-related shocks has further complicated the role of policymakers, especially in developing economies (OECD, 2017; United Nations, 2022).

In spite of the impact of COVID-19 on the advancement of the 2030 Agenda, the agenda still provides a suitable framework for recovery. The pandemic revealed the urgency to tackle the root causes of social and economic vulnerabilities in a coherent and integrated manner, and the pursuit of the 2030 Agenda by governments will maintain focus on crucial aspects for recovery: inclusion, equity, growth, resilience, and sustainability (Bettelli, January 2021; Mohieldin & Shehata, 2021). Consequently, this chapter evaluates the role of national governments in advancing the 2030 Agenda, in light of the COVID-19 pandemic developments, alongside three main factors that are necessary for accelerating progress toward the SDGs and "Building Forward Better." These elements consist of policy frameworks and their effective implementation, dependable data, and adequate finance. While this chapter analyzes each of these factors separately, it also highlights how the three of them work to reinforce one another.

Policy Frameworks and Effective Implementation

A common first step toward national governments' adoption of the 2030 Agenda has been establishing the structure of the governance system for the SDGs at the national level (Allen et al., 2018; Morita et al., 2020). Since 2015, countries have been undergoing changes in their institutional arrangements for engaging with the SDGs. Analysis of countries' VNRs, national development plans, and institutional arrangements show that the 2030 Agenda and the SDGs have gained visibility and political salience as an overarching policy agenda in both developed and developing countries. However, the process of institutionalizing the 2030 Agenda and incorporating it in national development policy frameworks is a work in progress that undergoes numerous stages. There is no uniform pattern that can be observed across countries for the institutional arrangements and policy frameworks introduced for the SDGs or for the speed of such institutionalization (UN DESA, 2021a). Furthermore, COVID-19 severely impacted many national institutions and threatened to decrease the political salience of the 2030 Agenda, reducing it to a secondary priority to address the urgent needs of the pandemic. However, both the

2030 Agenda and the pandemic call for the existence of national institutions that adopt policy frameworks that enhance policy coherence in all dimensions and work both horizontally and vertically. This is necessary to account for the multidimensional and interlinked social, environmental, and economic phenomena while balancing both short-run actions and the need for long-run investments (Mohieldin & Shehata, 2021; UN DESA, 2021a).

Policy Coherence and Horizontal Integration

One of the key lessons from the Millennium Development Goals (MDGs) is that even when goals are expressed as sector-specific, they cannot be achieved by adopting sectoral siloed policies (Martínez-Solimán & Mohieldin, 2016). Accordingly, the post-2015 Agenda (i.e., the 2030 Agenda) emphasizes the integrated and indivisible nature of the SDGs and that any progress requires the adoption of cross-sectoral objectives and implementation mechanisms. For instance, one perspective of the 17 integrated goals can be summarized as follows: within a framework of effective partnerships and good governance, policies and programs that enhance economic empowerment and investment in infrastructure in combination with adequate investment in human capital as well as resilience and environmental considerations, would eventually lead to fighting poverty and reduced inequalities (Mohieldin, May 2021). The 17 SDGs together emphasize progress on five critical themes: people, planet, prosperity, peace, and partnerships, thus adopting the principle of "leave no one behind." Indeed, this is an interconnected framework with a premise that one element affects the success of others, as opposed to a group of siloed goals. COVID-19 has also highlighted the need for policies that address what are often seen as divergent economic, social, and environmental priorities. What began as solely a health crisis went on to negatively impact communities and groups that were already lagging behind, caused significant economic and social consequences, and exposed how biodiversity loss and climate change continues to shape the present and future. Because the pandemic crisis is multifaceted, governments should design policies for the response and recovery phases using a coherent, multi-sectoral, and systematic approach instead of a sequential, siloed strategy (OECD, 2020a; UNDP, 2020a) (Fig. 2.1).

The ability to design coherent policies that capitalize on the interlinkages between the SDGs and minimize trade-offs requires both ex-ante and

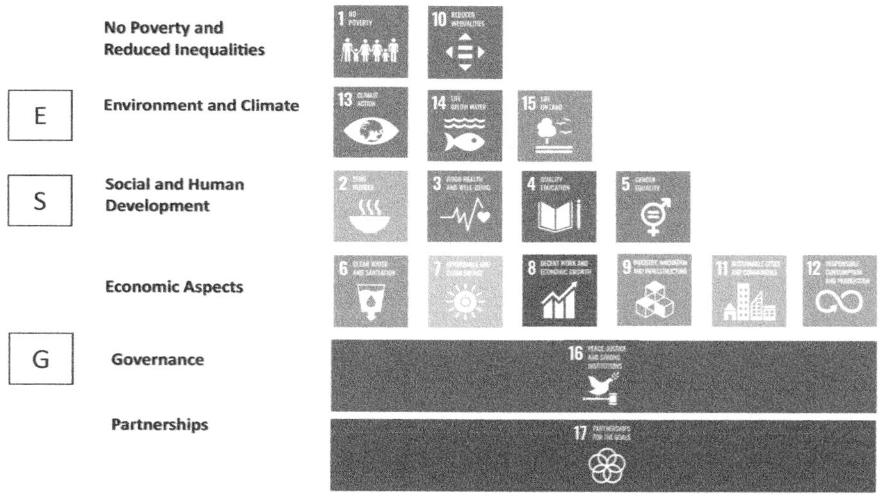

Fig. 2.1 The six integrated dimensions of the Sustainable Development Goals (*Source* Authors based on Mohieldin, M. (2021, May). *A global economic system in turmoil: What is at stake for the MENA region?* ERF)

ex-post assessment of proposed policies and programs. On this note, there are some emerging examples for sustainability assessments. Sustainability assessments represent systematic and iterative processes for the ex-ante assessment of the possible economic, social, and environmental effects of the proposed policies, plans, and projects. These assessments are typically carried out during the preparation phase and aim at improving the performance of the proposed policies by strengthening positive impacts while minimizing negative ones (OECD, 2008). Some of the OECD countries have thus recently attempted to adopt sustainability assessment tools with the aim of linking separate policy evaluations. For example, proposals for new laws and regulations in Germany undergo a sustainability impact assessment based on indicators, targets, and management tools that incorporate an intergenerational dimension. The introduction of the online tool eNAP, which allows laws and regulations to be evaluated against these indicators, serves to translate abstract regulations and clarifies the concept of sustainability by articulating concrete requirements (OECD, 2020a; Wanckel, 2021). Luxembourg adopted a sustainability check tool that became a formal part of the legislative cycle in 2020, in order to evaluate the effects of legislative acts in comparison to the government's sustainable development priorities. Ministries are required to answer questions on the impacts of the proposed laws on a set of 10 defined priorities and to explain how they intend to mitigate the potential negative spillovers (OECD, 2020a). However, despite the availability of these assessment tools, how widely adopted they are varies depending on the given laws and rules in place, as well as whether government departments are mandated to use them (Wanckel, 2021).

Despite the clear importance of cross-sectoral horizontal integration, evidence shows that the national implementation of the 2030 Agenda for Sustainable Development, and the Paris Agreement on Climate Change tend to advance on different tracks. Climate actions defined in the context of a country's Nationally Determined Contributions (NDCs) are usually defined and advanced separately from the national targets for the SDGs, resulting in a disconnected and siloed approach for addressing the interlinked challenges of climate change, increased poverty, and rising inequalities (Bouyé et al., 2018; Mohieldin, April 2021; Shawoo et al., 2020). This is further emphasized by the fact that mainstreaming the 2030 Agenda into national frameworks lacks reference to global agreements on climate change, disaster risk reduction, and the New Urban Agenda (Okitasari et al., 2019). A survey of 90 countries reveals that seven out of eight countries have different coordination mechanisms for both NDC and SDG implementation, and that policymakers report significant challenges in linking the two processes because each has 'its own history, community of actors, and political dynamics' (Bouyé et al., 2018).

The proliferation of different and disconnected guidance on the SDGs and NDCs impedes coordination efforts and overburdens the planning process. This lack of joint SDG and NDC processes problematically masks the sectors lagging behind where interventions are needed the most. Line ministries have the tendency to choose SDG and climate goals in a way that matches pre-determined sectoral targets, which results in missing to identify actions that may have capitalized on synergies and minimized trade-offs in a way that contributes to the advancement of both agendas. Governments need to recognize the concept of sustainability is not limited to climate change, and should work on designing policies that enhance the integration between the two agendas. Failing to do so may culminate in a two-tiered system; one that involves a fast track toward achieving the climate agenda and a slower track for the remaining SDGs (Mohieldin, March 2021). This would be a waste of the advantages that come from integrating the SDGs and neglect the opportunity of working toward a just transition, thus negatively affecting the SDGs' progress, particularly regarding jobs, poverty, and inequality.

The SDGs incorporate climate action across their goals and especially reference the role of the United Nations Framework Convention on Climate Change (UNFCCC) for defining climate ambition. In addition, the Paris Agreement requires that countries undertake climate actions in the overarching context of sustainable development and poverty eradication. Translating this integration into a national context calls for ensuring synergies and dialogue between the coordinating institutions for both agendas. This dialogue can occur through ensuring that the leading agencies of both agendas participate in national coordination bodies, and have in place common policy planning instruments to mainstream both agendas' policy objectives, as well as regularly cross-check their implementation strategies (Bouyé et al., 2018; Shawoo et al., 2020). Screening national climate actions to ensure cohesion with the rest of the SDGs and the overarching objective of "leave no one behind" provides a mechanism that efficiently designs and prioritizes policies that maximize synergies as well as directs benefits to the most vulnerable. Furthermore, climate action is expected to have notable effects on human development and economic growth, especially if aspects of a just transition are embedded in the policies' design. However, the significance of the other SDGs, given COVID-19, makes it imperative to transcend the concept of co-benefits and to act to design interventions that directly and concurrently advance progress toward the SDGs (Mohieldin & Shehata, 2021).

Similar to the gap in integrating climate action within the framework of the 2030 Agenda, the response to COVID-19 by many governments lacked horizontal policy integration. The multidimensional and multi-sector impact of the pandemic necessitated coordinated policy actions across a wide range of policy areas such as health, education, public transport, tourism, and social safety nets. This was necessary in order to control the spread of the pandemic, take measures to combat it, and implement the gradual lifting of these measures while supporting the segments that were most negatively impacted. One important area that exposed the need for policy integration in the context of COVID-19 and affected the progress toward the SDGs has been limiting and reversing the negative impacts of the pandemic and policy responses that exacerbated existing inequalities (UNDP, 2020a).

On the one hand, when focusing on the health effects of the pandemic as a result of limited resources, weak access to healthcare, and patterns of discrimination and disadvantages, it is clear that marginalized segments of the population have been the most vulnerable. But the pandemic also affected these groups on several levels including employment, education, and access to basic needs. Those with low-incomes are more likely to be occupying low-paying or informal jobs and to have experienced a huge disruption to their employment. Similarly, and as shown in the next section on dependable data, low-income groups tend to have less access to internet and thus experienced discontinuity in education as they could not effectively commit to online education systems (UNSDG, 2020). Tackling these compounded impacts of the pandemic calls for integrated policies that would also increase resilience to future shocks. Nevertheless, governments' response has been widely sectoral and temporary with no clear vision on how gradually these measures can be lifted without exacerbating inequalities (UN DESA, 2021b). For instance, as of September 2020, over 200 countries had introduced more than 1,000 social protection measures (Gentilini et al., 2020). According to the World Bank's February 2022 update, there was a total of 3,856 social protection and labor measures planned or implemented by 223 economies. However, the rate at which these measures were put in place around the world has since slowed down (Gentilini et al., 2020). Also, most of these measures were short-term and may not have taken into account some of the long-term effects of the pandemic, such as the closing of schools (UN DESA, 2021b).

Policy Coherence and Vertical Integration

Policy coherence also requires that national governments adopt governance mechanisms that support a whole-of-government approach. This should be done in order to foster vertical integration as well as multi-stakeholder partnerships with a wide range of actors that go beyond the notion of consultations (Okitasari et al., 2019). Vertical integration entails consistency, coordination, and collaboration across different levels of government. The 2030 Agenda emphasizes the importance of local governance in achieving the SDGs, as an estimated 65% of the 169 targets may not be met without proper engagement with LRGs (SDSN, 2016). LRGs are also key for implementation at the last mile to ensure that "no one is left behind." COVID-19 has both emphasized the importance of LRGs, who have been at the frontline in fighting the pandemic, as well as exposed the challenges of vertical coordination to ensure coherent response measures. In all cases, the depth of the role played by LRGs in the development agenda is largely shaped by the governance mechanisms adopted by national governments and their commitment to the localization process.

It is first necessary to acknowledge the territorial nature of the SDGs. Concentrating solely on thematic areas is an insufficient means of SDG policy implementation, as targeting the last mile requires incorporating territorial specificities in all designed policies and programs. Investing directly in the last mile where subnational governments interact with communities ensures the comprehension and handling of exclusionary factors as opposed to depending on a top-average view of development that tends to mask territorial heterogeneities, which would further exclude those lagging in progress (European Commission, 2018; UNDP & UNCDF, 2016). It is worth noting that the capacity of LRGs to act on last-mile delivery is largely determined by national governments' commitment to mobilize them (UCLG, 2017, 2019). As such, there has been a high participation of LRGs in many countries where there is a present national commitment to include all governance levels. This has taken place through various mechanisms, including promoting outreach aimed at local governments as seen in the case of Finland, the fostering of policy alignment between the SDGs and local development plans in Colombia and South Africa, and the involvement of LRGs in monitoring and follow-up in Belgium, Sweden, and Denmark. Furthermore, the actions of national governments gain significance in centralized systems such as that of Jordan, when the state adopted a strategy to promote the ownership of SDGs at the subnational level (UCLG, 2017). Localization has long been cited as a positive way to involve subnational governments in aligning the SDGs with local development priorities. One such notable instance of national-local collaboration is exemplified in Colombia, where local authorities and departmental capital cities have national support to adopt local development plans that include localized SDG targets. However, the reality is that in many cases national governments view the localization process as a top-down relationship where the SDGs are passively shifted down to LRGs. This approach risks portraying a misconception of the SDGs as an external burden that works to hinder local participation, and foregoes last-mile implementation (UCLG, 2017, 2019).

COVID-19 further highlighted the need for vertical integration and revealed how there is no concrete answer regarding what constitutes an optimal degree of decentralization for crisis management. On the one hand, a fully decentralized approach may cause subnational and local governments to compete, as evident during the first months of the pandemic. On the other hand, highly centralized decision-making that ignores consultations with subnational levels of governments can lead to confusion surrounding rule application, especially given the territorial and heterogeneous nature of the pandemic. This has the potential to incite major social issues for local governments, as was the case when informal workers were forced to leave cities they could no longer work in (UN DESA, 2021b). Another point of importance is maintaining a balance between the perceived need for coordinated actions across all levels of government and sustaining some degree of flexibility in local responses, which as seen in Germany, is a dynamic process (OECD, 2020b; UN DESA, 2021b). It can be said that the most significant governance aspect COVID-19 developments emphasized is that a coordinated crisis response throughout all government levels is essential in minimizing crisis-management failures as well as avoiding disjointed or duplicated responses (OECD, 2020b).

Overall, the pandemic is expected to reshape the future of multi-level governance. More than 70% of the respondents from the June 2020

European Committee of the Regions (CoR)-OECD survey of 300 EU regions and municipalities indicate that the lack of coordination with other levels of government was among the biggest challenges faced in managing the crisis, alongside a lack of technical means and limited financial resources. As such, clearly established roles and responsibilities among levels of government were indicated as a high priority area where national government support is needed, especially when adopting recovery measures. Coordination is thus the highest priority given that it was the most cited factor by respondents in managing a successful exit strategy, particularly in terms of the design and implementation of measures across levels of governments, with the issue of additional financial resources coming in second (OECD, 2020c).

Attracting the Private Sector

Beyond vertical coordination, an important partner in advancing the 2030 Agenda is the private sector. A private sector that invests in the future would greatly benefit a country in meeting their development goals. This especially applies to investments in the areas of productive capacity and infrastructure development when it is implemented in an inclusive and deeply sustainable manner (United Nations, 2022). A shared element between LRGs and the private sector is that the role of each in the development agenda, including implementation at the last mile, is largely shaped by the governance mechanisms, laws, regulations, and policies that national governments adopt. For instance, states can utilize investment policies to attract private investment in the SDGs by directing investment promotion and facilitating the adoption of necessary risk mitigation instruments to address the shortcomings of the investment environment. This is supported by how governments in more than 150 countries have developed national sustainable development strategies or revised existing development plans to encompass the SDGs, with over 70 countries in the process of establishing their integrated national financing frameworks (INFF) to explore the financing of these plans. Such actions impart a strong signal to the private sector in an attempt to attract their contributions in a manner that aligns with state priorities (United Nations, 2022; World Bank, 2021a).

The mobilization of the private sector must be carried out with sustainability embedded in core operations. The private sector's role in the achievement of the SDGs will continue to remain unclear without a transparent view of their impact. Transparency regarding the corporate disclosure of sustainability-related information is hindered and unnecessarily complicated by the existence of fragmented reporting frameworks and the multiplicity of surveys and questionnaires that companies are asked to fill. Besides the issue of harmonization, defining the scope of sustainability reporting in terms of single vs. double materiality is necessary. The conservative view requires companies to disclose sustainability risk information that impacts their financial performance. A more comprehensive framework would have information disclosed on the external impact of the company's operations on social and environmental issues. In all cases, the necessary harmonization and scope definition of company sustainability information disclosure calls for leadership from policymakers. For example, the International Financial Reporting Standards (IFRS) became the primary financial accounting standards in more than 140 jurisdictions after being adopted by their national regulators, and a similar agreement is required for sustainability reporting. Defining the scope of sustainability reporting calls for establishing mechanisms at the national level to allow for coordination between different stakeholders which would pave the way for adopting a common framework (United Nations, 2021a, 2022). This should be complemented by making reporting mandatory to allow for comparability. Many regulators are moving toward mandating corporate reporting, but the move has for the most part focused on climate-related issues, thus contributing to the two-tier system previously discussed (TCFD, 2021; United Nations, 2022). Moreover, since most sustainability disclosure regulations apply only to listed companies, this may risk public companies selling their carbon-intensive assets to others that do not have the same transparency requirements. This is problematic given that numbers show in the past couple of years, private equity funds acquired USD 60 billion worth of oil, gas, and coal assets: a third more than their investment in renewables (*The Economist*, February 2022).

Corporate information disclosure on sustainability can be complemented by regulations that mandate fund managers to disclose the environmental and social impact of their clients' portfolios, and others that require employment retirement plans to have an investment alternative focused on achieving a positive impact on sustainable development. Regarding the latter, one early step is being taken by the United States Department of Labor, which has proposed to ease the process for employers to provide retirement options that factor ESG considerations in investment decisions (United States Department of Labor, October 2021). On the matter of transparent asset allocation and investment decisions of institutions that manage funds, it is still at the discretion of fund managers to release sustainability information about such funds. Nevertheless, there are developing regulations such as the European Union Sustainable Finance Disclosure Regulation (SFDR) that will require such disclosure (United Nations, 2022). These regulations are key in capitalizing on individual investors' interests regarding the impact of their investments on the SDGs. An example of such interest lies in the Netherlands, where two-thirds of pension fund respondents indicated their willingness to expand the fund's engagement with companies based on the SDGs, regardless of the effect on financial performance (Bauer et al., 2021).

Balancing the Short Term and the Long Term

At the heart of the 2030 Agenda is the need for long-term investments to close development gaps. Consequently, governments' commitment to the SDGs is often complicated by the pressure exerted from immediate economic and social pressures, with short-term goals prioritized by the nature of electoral terms (de Mello, 2016). The pandemic has further exacerbated the short-term and long-term policy dichotomy, with the economic consequences of the pandemic believed to have side-lined long-term sustainable development policies and well-being priorities (OECD, 2020a; UN DESA, 2021a). For instance, as of February 2021, while 35 countries had national recovery plans, less than half of them referenced the SDGs, and if so, it was in the general narrative and not as the main pillar to guide a sustainable, inclusive, and resilient recovery (Sachs et al., 2021). This was done despite the need for a sustainable recovery built on policy coherence in order to balance short-term responses to the pandemic with managing investments, as well as inducing behavioral changes to minimize vulnerabilities and the probability of future shocks. A sustainable recovery is also one that demands coordinating the needs of the present-day population with those of future generations. By adopting short-sighted measures that stimulate the economy at the expense of environmental and social costs, governments will miss out on the 'double dividends' of sustainable recovery (OECD, 2020a).

Building Forward Better, for one, requires resilience and investments in infrastructure and human capital (Mohieldin & Shehata, 2021). The pandemic has exposed how many health systems lack preparation for crises and have a need to upgrade their infrastructure in terms of health and inclusivity. Although public infrastructure is a key driver of inclusive economic growth and development as well as helpful in building resilience against shocks, a well-planned and implemented infrastructure is not an easy target. Public investment is often plagued with losses, with estimates showing that on average countries lose more than one-third of their resources spent on establishing and maintaining infrastructure as a result of inefficiencies, particularly in connection to poor infrastructure governance. No income country group is immune to the efficiency gap in public infrastructure. The efficiency gap that results from poor infrastructure governance reached 53% in low-income developing countries, 34% in emerging markets, and 15% in advanced economies (Baum et al., 2020). Infrastructure governance refers to the 'institutions and frameworks for planning, allocating, and implementing infrastructure investment spending' (Schwartz et al., 2020, p. 1). Also, optimism bias in project appraisal and planning results in cost overruns and poorer than projected benefits (Flyvbjerg, 2009). Corruption, weak coordination among government agencies, the selection of projects based on political considerations and challenges in project implementation all contribute to inefficient public infrastructure investment (Rajaram et al., 2014). However, stronger institutions can play a major role in potentially closing the efficiency gap by more than half (Baum et al., 2020). Moreover, countries with high efficiency in public investment are less likely to witness spending cuts during periods of fiscal constraints, which conveys that good infrastructure governance is important to help countries reconcile fiscal sustainability considerations with the preservation of public investment (Basdevant et al., 2020).

Infrastructure needs to be resilient in order to withstand shocks and disasters. Furthermore, by incorporating resilience into infrastructure investment, significant damage and associated output loss in public infrastructure can be avoided (Guo & Quayyum, 2020). Nevertheless, the matter of investing in resilience goes beyond infrastructure. A risk and resilience framework is one that considers the systemic nature of risk and requires complementing the traditional risk management frameworks with risk reduction and resilience instruments. For instance, a risk/resilience framework incorporates efforts to both reduce and manage risk by complementing the risk assessment and risk management phases with resilience strategies that target strengthening a systems capability to maintain basic functionality during crises, as well how to better adapt to new conditions (United Nations, 2021a). Governments have yet to include risk analysis into their planning processes in order to build resilience into public budgets, as well as avoid ex-post biases in implementation and funding. Countries have the tendency to allocate more funds for crisis response than to ex-ante efforts of risk reduction, which is fueled by the limited understanding of the risks landscape and alternatives for risk reduction measures, as well as the high political visibility of ex-post efforts (United Nations, 2021a). Governments can conduct multi-hazard risk assessments to better understand the risk landscape and its potential effects. They can also benefit from strategic foresight tools to undertake decision-making in highly uncertain environments. Strategic foresight processes include a variety of measures. These include horizon screening to research signals of change to the status-quo and analyze their potential impacts, megatrends analysis to review large-scale and multidimensional phenomena, scenario planning by forecasting how the future looks in different scenarios; as well as visioning and back-casting (which involves developing a future state and identifying the challenges and necessary steps of reaching that end) (OECD, 2020a, 2020d).

Lastly, a stable and lasting recovery calls for restoring human capital while strengthening service delivery systems that can build, preserve, and utilize this capital. The pandemic caused significant damages to human capital that must be addressed with well-planned and sufficient investments, in order to avoid the persisting detriments of COVID-19 on future productivity (UNDP, 2020a; World Bank, 2021a). For example, the pandemic caused significant learning losses, with estimates claiming that the percentage of 10-year olds in low and middle-income countries who cannot read a basic text has increased by 17 percentage points between 2019 and 2021, reaching a peak of 70% (World Bank, October 2021). This depth of learning poverty combined with the deskilling effects from prolonged periods of unemployment and underemployment are expected to reduce future productivity and earnings (World Bank, 2021a). Thus, these effects call for adequate, efficient, and sustainable investment spending toward re-strengthening human capital, especially during periods of fiscal tightening. These investments are also critical for a green, inclusive, and resilient recovery. Countries should work to balance physical infrastructure spending with human capital investment as the former tends to be prioritized because of its visibility, and while the benefits of human capital investments tend to accumulate slower and are more difficult to measure (Mohieldin, 2017; World Bank, August 2018).

Dependable Data

The SDGs are quantitative by design and are backed by a long list of targets and indicators that necessitate the presence of high-quality, granular, and timely data. However, the role of data in the 2030 Agenda surpasses the progress monitoring function. The Addis Ababa Action Agenda (AAAA) stresses the importance of high-quality disaggregated data for smart, informed, and transparent decision-making that strengthens the means of implementation and aims to improve policy-making processes at all levels (United Nations, 2015b). To meet the pledges and ambitions of the SDGs, the Paris Agreement, and the Sendai Framework for Disaster Risk Reduction, governments need to understand people's vulnerabilities, evaluate services and programs, and anticipate future opportunities, risks, and trends. This, in turn, emphasizes that governments need three types of data: timely data to gauge the efficiency of policies and intervene in a timely manner, historical data to understand past trends and capitalize on the past lessons, and forward-looking modeled data to better prepare for the future (SDSN TReNDS, 2017). The pandemic further demonstrated the significance of timely and accurate data on the population, health, and the economy at large. Many countries have harnessed data to control the pandemic, as seen in the example of tracking people's locations to capture mobility patterns during lockdowns and trace disease contact (United Nations, 2022; World Bank, 2021b). Despite the fact that national statistical systems are expected to play a key role in generating, disseminating, and administering data, information from other stakeholders such as the private sector and civil society should also be utilized to achieve development outcomes.

Data Collection and Measurement

The response to COVID-19 has required timely and disaggregated data that allows for the assessment of vulnerabilities to administer short- and long-term recovery measures. The SDG indicators by nature provide a holistic view of the impact of the pandemic on the health, social, and economic fronts for different age groups and the most vulnerable. In addition, statistical authorities have used SDG indicators to describe the impact of COVID-19 (OECD, 2020a). The pandemic also resulted in the introduction of innovative approaches to address the increased data needs. Several National Statistical Offices (NSOs) used alternative methods of data collection, such as telephone or web surveys, as well as depended on alternative data sources such as geospatial and administrative data (United Nations, 2022). In some cases, this was coupled with an increase in experimental data collection to respond to governments' need to have timely knowledge on the movements of patients so that support could be provided. For example, the Dutch Central Bureau of Statistics (CBS) in the Netherlands had a knowledge and coordination point set up where private and public data owners shared data with the CBS. These owners include the Dutch Payments association, retail scanners, traffic loops, payment systems, and the airport. The collected data was then combined with administrative data that was used to support policies such as tracking applications in support of self-employed professionals under the Temporary bridging measure for self-employed professionals (Tozo) (OECD, 2020a).

However, national data systems were for the most part unprepared to respond to the pandemic-induced demand spike for data, thus exacerbating global data inequalities that existed before the pandemic, especially in developing countries (United Nations, 2022). According to a survey by the United Nations and the World Bank, in sub-Saharan Africa, 97% of the surveyed countries responded that the production of regular statistics was affected by the pandemic. Results also showed that 9 in 10 NSOs in low and lower-middle-income countries witnessed funding cuts and faced difficulties in operations (United Nations, 2020a). The effect of the pandemic on NSOs has been more extensive and prolonged than anticipated when after one year into the pandemic statistical offices were still facing challenges in resuming their operations. Surveys and census were the most hard-hit operations as 81% of NSOs in low and lower-middle-income countries report delays in their surveys compared to 61% in high-income countries, further highlighting the divide in countries' ability to cope and adapt (Hammer et al., 2021). Moreover, the pandemic has disrupted data collection on population groups that are expected to be among the most negatively affected, thus limiting governments' capacities to design responses to alleviate the effect of the pandemic. Migrants, internally displaced people, and people with disabilities have been identified as difficult-to-reach populations by statistical offices, with limited phone and internet coverage cited as the primary causes of reachability issues (Hammer et al., August 2021).

Global data inequalities are further reflected in the gap between developed and developing nations' performance on the Statistical Performance Indicator (SPI), which measures statistical performance across five pillars and a total of 22 dimensions. These pillars are data use, data services, data products, data sources, and data infrastructure (Dang et al., 2021). Developing countries, despite some improvements prior to COVID-19, scored much less than developed countries, especially on data sources and infrastructure pillars. This particularly applies to Small Island Developing States (SIDS) and Least Developed Countries (LDCs). A closer look at the performance of the top performers on the SPI reveals that significant gaps remain on pillar 3 (data products), which measures the ability to produce relevant indicators, especially for the SDGs (Sachs et al., 2021). Despite the fact that all 231 official SDG indicators had internationally established methodologies as of 2021, they fell into the Tier II category when more than 40% of these indicators had yet to be regularly produced by countries (Sachs et al., 2021; UNSD, 2021). Furthermore, there are marked differences across the SDGs in terms of data availability and timeliness (Sachs et al., 2021; United Nations, 2020a). Specifically, SDG 13 on Climate Action and SDG 14 on Life Below Water encounter the dual problem of limited coverage and time lags in reporting, highlighting the substantial gaps for environmental SDG indicators (Dahmm, April 2021; Sachs et al., 2021). These gaps convey how despite recent developments in SDG reporting, current progress is insufficient, with governments missing the opportunity to acquire the necessary information on the environmentally, socially, and economically affected groups and regions. Although the percentage of potential data points to be filled every year jumped from 43.6% in March 2019 to 58.3% in March 2021, it remains true that only half of the data needs are being met (Kitzmueller et al., August 2021; Sachs et al., 2021) (Fig. 2.2).

Another aspect of data equations is the importance of having quality data on the private sector's contribution to the SDGs so that governments can monitor national progress and design policies to incentivize and potentially enforce sustainable practices. Companies exert environmental and societal impacts through various channels, including through carbon emissions and labor practices. Large companies are increasingly publishing sustainability reports, as seen through how in 2020, approximately 92% of S&P 500 companies published a sustainability report

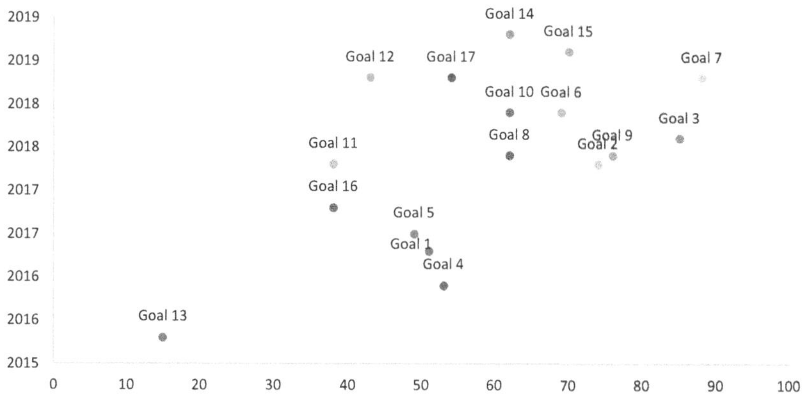

Fig. 2.2 Data Coverage and timeliness for Official SDG Indicators (2021) (*Note* Data timelines as reflected on the vertical axis shows the most recent year for which data on the Goal is available, as a weighted average of its indicators. Data coverage captured on the horizontal axis shows the percentage of countries or areas that have available data on the Goal. *Source* Author, based on United Nations [2021b])

compared to only 11% in 2011 (United Nations, 2022). Nevertheless, the quality of the information provided needs to be enhanced. This is because corporate sustainability reporting suffers from a number of issues that limit the ability to understand the environmental and social footprint of these companies. Information provided does not usually allow for comparability across companies and timeframes, and in many cases this information is based on qualitative indicators. Due to the voluntary nature of sustainability reporting, companies may choose the issues to communicate in these reports. Furthermore, such reports still lack independent assurance. All of these factors reduce the reliability and quality of the disclosed data (United Nations, 2021a, 2022).

Granular, detailed, and timely data is needed to build measures that capture the multidimensional aspects of progress (Stiglitz et al., 2018). The varied effects of COVID-19 and the mounting climate crisis emphasize the necessity of developing measures that go beyond GDP. GDP has long been used as a measure of economic prosperity, with GDP per capita adopted as a measure of average standards of living and economic well-being. Despite its wide usage, there have been many concerns over the limited scope and inadequacy of the GDP as a measurement. For example, it does not incorporate dimensions of well-being, distribution, the effect of digitalization, non-market activities, or economic and environmental sustainability. Limitations of the GDP were discussed by the Commission on the Measurement of Economic Performance and Social Progress (Stiglitz-Sen-Fitoussi Commission) which demanded that more attention be given to indicators of income, consumption, and wealth (Stiglitz et al., 2009). The GDP limitations are extended to a lacking reflection of the full effect of the pandemic. For instance, it does not adequately capture the effect of COVID-19 on the care and informal economy, nor does it accurately convey the location-specific and differentiated impacts of the pandemic. Currently, much work is being carried out to update the 2008 System of National Accounts. To elaborate, one of the main areas of this work set to be completed by 2025 is to tackle the issues pertaining to well-being and sustainability and define a broader framework of accounts to enhance monitoring and analysis capacities. The main approach adopted by the Intersecretariat Working Group on National Accounts (ISWGNA) is to have additional extended accounts linked and feeding into the broader System of National Accounts. These extended accounts span the following areas: distribution of household income, consumption, saving and wealth, unpaid household service work, education, and human capital, health and social conditions, and environmental-economic accounts (United Nations, 2020b, 2022).

Data Governance Framework

The availability of timely, disaggregated, and comprehensive data largely depends on the components of the national data system. A national data system transcends the boundaries of the data and statistics produced by official data bodies to include the data produced, exchanged, and used by participants from the public and private sectors. According to the World Development Indicator Report 2021: *Data for Better Lives*, a national data system is built on the pillars of infrastructure policies, laws, and regulations, economic policies, and institutions needed to effectively govern data. They represent the layers of the data governance framework that should allow countries to realize the full economic and social value of public and private intent data as well as enhance the synergies between them (World Bank, 2021b). These pillars need to be based on the foundations of human capital, trust, funding, incentives, and data demand. Ideally, a national data system should be backed by a national data strategy that outlines the steps needed to design and implement the pillars of the national data system while taking into consideration the country's data maturity stage. Data strategy usually serves as guidance regarding the needed support to strengthen the data system, and windows of cooperation with third parties looking to provide such support (SDSN TReNDS, 2017; United Nations, 2022; World Bank, 2021b). Yet, despite the importance of data strategy, not many countries have adopted a stand-alone one. For high-income countries, only 52% have this strategy compared to 18% of upper-middle-income countries, 6% of lower-middle-income countries, and a mere 3% for low-income countries (World Bank, 2021b) (Fig. 2.3).

Despite this limited application of data strategies, it is still necessary to investigate the pillars of the national data system to understand issues pertaining to data governance. These pillars can be viewed as layers that build on one another, with the first layer being the policy framework for data infrastructure. This involves both the promotion of access to data services and policies covering the infrastructure needed for exchanging, storing, and processing of data (World Bank, 2021b). The richness of the data and its social and economic value are directly related to the number of citizens connected, though universal access is not an easy target. Within the range of broadband signal, approximately 40% of the world's population lack access to the internet, conveying how policies should also tackle demand-side barriers that limit service uptake. These barriers include limited digital literacy, affordability, and relevance, thus exacerbating the digital divide (Chen, 2021; World Bank, 2021b). Additionally, there is a clear digital gender divide as statistics show that there are 250 million fewer women worldwide who use the internet compared to men (World Bank, 2021b). These factors imply that governments have yet to work on increasing the digital knowledge of certain segments of the population when partnerships with local governments and CSOs can address this issue. Governments can also work to ensure that primary education curricula are designed to equip the students with the necessary digital skills and knowledge. On the affordability barrier, there are already efforts to reduce the entry-level cost of internet and mobile devices (World Bank, 2021b). Massive data acquisition about population needs to be complemented with adequate data infrastructure to store it. National data platforms or portals are particularly important for official statistics and databases. Nevertheless, the presence of these data portals and their

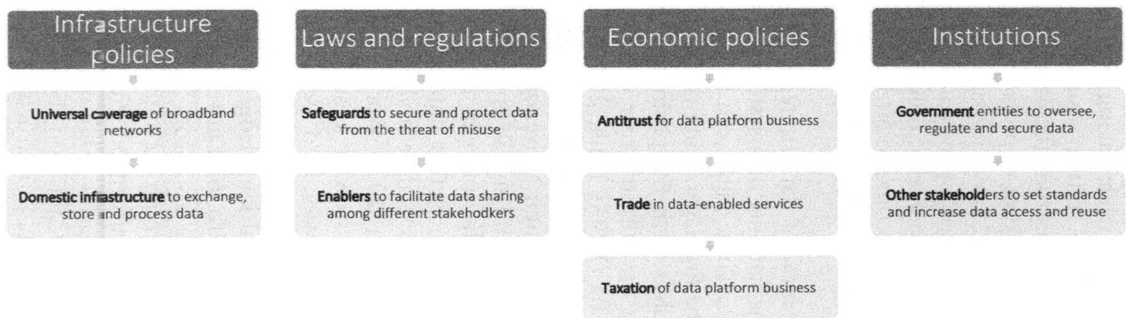

Fig. 2.3 Layers of data governance at the national level (*Source* Adapted by the authors based on the World Bank [2021b])

level of advancement varies among income country groups (ODIN, 2021; United Nations, 2022; World Bank, 2021a).

The second layer of the data governance focuses on the legal and regulatory environment, with the rules that both enable and govern the reuse and sharing of data while protecting against possible data misuse. This legal and regulatory framework for data transactions is a prerequisite for instilling trust in the transactions and in most countries the development of such a framework remains a work in progress (World Bank, 2021b). The first set of rules is the safeguards aiming at minimizing the harm from the misuse of data. Cybersecurity is the first fundamental foundation for these safeguards (ITU et al., 2018). A notable challenge in introducing a cybersecurity legal framework is that there does not exist a universally accepted definition for cybercrime and that it is not confined by borders (World Bank & United Nations, 2017). Only 40% of high-income countries in the Global Data Regulation Survey have security requirements that data controllers and producers should abide with. Despite the limited number of lower-middle-income countries with these security requirements, Kenya is cited for having comprehensive cybersecurity requirements in its Data Protection act (World Bank, 2021b). The data legal framework should also address the issue of data protection that requires a clear differentiation between personal and nonpersonal data. Unlike cybersecurity requirements, data protection laws are more common, as 137 out of 194 countries[1] have data protection laws (UNCTAD, 2022). However, the quality of the personal data protection laws in place seems to be a problem, especially in low- and middle-income countries (World Bank, 2021b). One notable development in this area is that the emergence of innovative analytical techniques and algorithm-based technologies like big data is blurring the line between personal and nonpersonal data. This requires that regulators establish frameworks to deal with this emerging mixed data.

Realizing the full value of data further necessitates the presence of enablers that allow for the reuse and sharing of data. The mechanisms by which governments and regulators support data sharing differs across public and private intent data, which reflects the degree of control that governments possess over these data types (World Bank, 2021b). Unlike public intent data which governments have control over, most private sector data transactions are usually based on voluntary contractual agreements. Thus, governments largely have to create incentives that would promote private data sharing. Still, in many cases, some governments have introduced legislation mandating the private sector to share data deemed to be in the public interest (OECD, 2019a). The growing interest in private intent data emanates from its unique features as it allows for constant collection of data in real time, it usually has a location aspect and it is less subject to bias information. Repurposing private intent data and combining it with that of public intent became of great relevance in the realm of tackling crisis and development issues (United Nations, 2022; World Bank, 2021b). However, in light of the previously discussed digital divide, private intent data might be skewed toward wealthier and more educated segments of the population who tend to use digital technologies more, as private intent data is often the by-product of the use of digital technologies (United Nations, 2022).

These safeguards and enablers for data interact with the wider economic policy issues and affect the country's ability to harness the economic value of data through taxation, trade, and competition. For example, the degree of stringency of data safeguards defines the pattern for cross-border trading related to data-driven services, and enablers that encourage data sharing among market players tend to affect competition. The interacting nature of these economic areas complicates the policy scene for policymakers who have to develop agile competition policies and modern trade and tax administrations that need to be transboundary. This, in turn, calls for strong institutional capacity and a capable institutional ecosystem that ensures the maximization of the social and economic value of data while effectively enforcing the data laws, regulations, and policies (World Bank, 2021b). To effectively carry out these functions, institutions ought to be formally mandated, equipped with the technical capacity to undertake the mandated functions, and have sufficient resources, especially financial ones (World Bank, 2021b). These three characteristics overlap with some of the foundations upon which the data pillars of the data governance framework are anchored, specifically regarding funding and human capital.

The availability of sufficient and stable financial resources is a prerequisite for a well-functioning national data system. However, data systems have long been underfunded. The issue of underfunding of national data systems can be analyzed across three dimensions: underinvestment by governments, underinvestment by donors, and fragmentation and lack of coordination (SDSN TReNDS, 2017; United Nations, 2022; World Bank, 2021b). Despite the formal expression by governments on the importance of data, especially as expressed in the VNRs, investments in data are not seen as a political imperative and the use of data is not yet viewed as fundamental in creating public value (OECD, 2019b; SDSN TReNDS, 2017). This coupled with the limited fiscal space available for many countries has led to governments underinvesting in their data systems, thus exacerbating the data inequalities among countries. Among high-income countries, 93% have a fully funded national statistical system compared to 0% for low-income countries (World Bank, 2021b). The COVID-19 pandemic has further worsened the scene particularly with the increase of data collection costs contrasting with a significant decrease between 2019 and 2020 in the number of countries that reported the availability of government funding (PARIS21, 2021; United Nations, 2022). Like governments, donors tend to underinvest in data. Global donor commitments to data and statistics have stayed almost stagnant since 2015 and are expected to have decreased with the pandemic. Over half of the sub-Saharan African countries reported a decrease in funding from donors since the start of the pandemic (World Bank, UNSD & PARIS21, 2022). But even when there are investments by donors in data systems, there is a misalignment of incentives and lack of coordination between the flows and the priorities of the national system. Donors' support is for the large part skewed toward short-term investments rather than long-term ones dedicated to strengthening data systems such as enhancing technical capacity, research and development, and infrastructure (OECD, 2021). This fragmentation in data systems funding calls for government coordination and ownership to recognize their data systems' needs and align funds with national priorities to ensure sustainable results. This can be reflected in a national data plan, if it exists, or in regional and subregional programs with partners.

The underfunding dilemma is feeding into the human capital needs of the data system, especially the capacities of data producers and institutions. The insufficiency of the funds directed toward long-term investment in technical capacity results in limited technical capacity and the inability of national data offices to attract the skilled data personnel and statisticians, which constrains the performance of public data producers. Securing skilled human resources or strengthening the capacity of the ones already on-board has been cited by more than 80% of African NSOs as one of their five most important targets for capacity development (PARIS21, 2018). Nevertheless, recruitment and retention of skilled personnel usually require competitive pay scales and progressive career tracks, which are often offered by private data businesses (Dargent et al., 2020).

Adequate Finance

The discussion up to this point portrays that progress toward the SDGs hinges on the availability of adequate finance. Finance goes beyond mere funding, as spending more does not necessarily warrant better outcomes. This note is crucial given that according to all measures, the level of financing required to achieve the 2030 Agenda was high even before COVID-19, after which the issue only worsened. Initially, the global investment needs for the SDGs were an estimated USD 5–7 trillion per year, with investment needs in developing countries alone ranging between USD 3.3–4.5 trillion per year for mainly basic infrastructure, food security, and climate mitigation and adaptation (UNCTAD, 2014). Focusing on the investments needed in the areas of education, health, roads, electricity, water, and sanitation, it is projected that the average additional spending needed for low-income developing countries represents 15 percentage points of their GDP in 2030, while these percentage points for emerging economies is 4.2 (Gasper et al., 2019). Overall, developing countries faced an annual financing gap of USD 2.5 trillion in the pre-pandemic era (UNCTAD, 2014). After the pandemic, the financing gap for developing countries spiked to USD 4.2 trillion in 2020, which was fueled by USD 1 trillion of additional needs for COVID-19 and a USD 700 billion drop in external private sources (OECD, 2020e). Furthermore, this gap comes alongside the asymmetric distribution of investments among different country groups. For instance, LICs have 8.4% of the world's population but less than 1% of the world's investment and LMICs account for 42.9% of the world's population but have only 15% of investment spending. In contrast, HICs comprise 15.8% of the world's population while accounting for almost half of the world's investment spending (IMF, 2019; Sachs et al., 2022).

However, if one considers the global public and private financial resources, global public and private investment would be sufficient in closing the SDG financing gap under the condition that these financial resources are invested in and aligned with sustainable development. For instance, public policies and the mobilization of domestic resources coupled with efficient and effective spending aligned with the SDGs is crucial to the pursuit of sustainable development through the domestic public resources channel. The ambitions of the 2030 Agenda would exert significant pressures on public budgets and capacities and thus require scaling up and increasing the effectiveness of external support, through both concessional and non-concessional financing. These domestic and international public resources also need to unlock domestic and international private investments and better align such investments with public goals, which in turn calls for countries to strengthen their enabling environments. In all cases, it is inevitable that countries resort to borrowing to finance their SDG aspirations. However, this needs to be based on a prudent debt management strategy within an international framework that allows for attaining long-term debt sustainability, which can be done through coordinated policies aimed at fostering debt financing, relief, and restructuring. Aligning these financial flows with the country's national development plans necessitates having a framework that governments can use to strengthen the integration between long-term sustainable development aspirations and the policies that would mobilize the investments needed to realize them. The Integrated National Financing Framework (INFF) assists governments in managing a complex financing landscape given a set of national development priorities by increasing coherence among financing sources and acknowledging the interrelated nature of climate, sustainability, and inclusion (UNDP, 2020b). While it is acknowledged that all these channels are necessary, public finance and public financial management remain central to the pursuit of sustainable development. Fiscal policies, the effective use of debt proceeds, and reflecting national sustainable development priorities into the state budget can all work to unlock greater financial flows toward SDG investments.

The Varied Capacity for Crisis Response

Differences in the fiscal space and policy tools available to the developed and developing countries resulted in considerable inequalities across these two groups. The magnitude of fiscal measures reached around 10% or more of GDP in the most developed nations, with a recorded high of over 20% in some large economies (United Nations, 2022). In contrast, the fiscal response of low-income countries was uniformly small, and notably varied among middle-income countries. This is owed to significant differences in the capability and willingness of governments to mobilize and direct fiscal resources toward support programs (World Bank, 2022).

The responses of developing countries were constrained by declining tax revenues that reflected tightened economic activity. In many cases, developing countries were left with no choice but to cut critical public investments to sustainable development such as in education, and reallocate resources to fund crisis response initiatives (United Nations, 2022; World Bank, 2022). This contrasts with the sufficient fiscal space available to many developed countries which allows them to both take the necessary immediate response measures, as well as channel resources toward strengthening social protection and bolstering productive investments in areas of digital technologies, research and development, and green energy (United Nations, 2022). Examples of these large-scale plans include the European Union's "NextGenerationEU" recovery plan, and the United States' "Infrastructure Investment and Jobs Act" (Fig. 2.4).

Despite the fact that many emerging economies already had high levels of sovereign debt before the pandemic, additional debt was taken on to make space for tackling the economic and human costs of the pandemic. Consequently, COVID-19 caused a surge in sovereign debt that is unusual in terms of both its speed and synchronicity. Moreover, around 60% of least developed and other low-income countries are already in debt distress or at high risk, while about a quarter of middle-income countries remain at high risk (United Nations, 2022; World Bank, 2022). Many developing countries were not able to benefit from the low levels of interest rates that existed during the pandemic in order to finance their investments, and sovereign borrowing costs for them remained much higher than that of developed states. Debt servicing is representing a greater risk for low and middle-income countries as result of the elevated cost of debt compounded by limited revenues. For example, in over half of sub-Saharan countries, debt servicing costs consume at least one-quarter of the government's revenue (United Nations, 2022). Moreover, the G20's Debt Service Suspension Initiative (DSSI) expired at the end of 2021, implying that countries are required to pay both present and previous debt servicing costs. These costs are directing further resources away from crisis response measures and investments in sustainable recovery. With mounting risks to debt sustainability, the majority of developing countries have started withdrawing most stimulus measures or are expected to do so soon.

However, given the uncertain global outlook, premature withdrawal of these fiscal stimulus packages can result in halting the recovery process and leading to higher debt-to-GDP ratios. This sovereign debt outlook

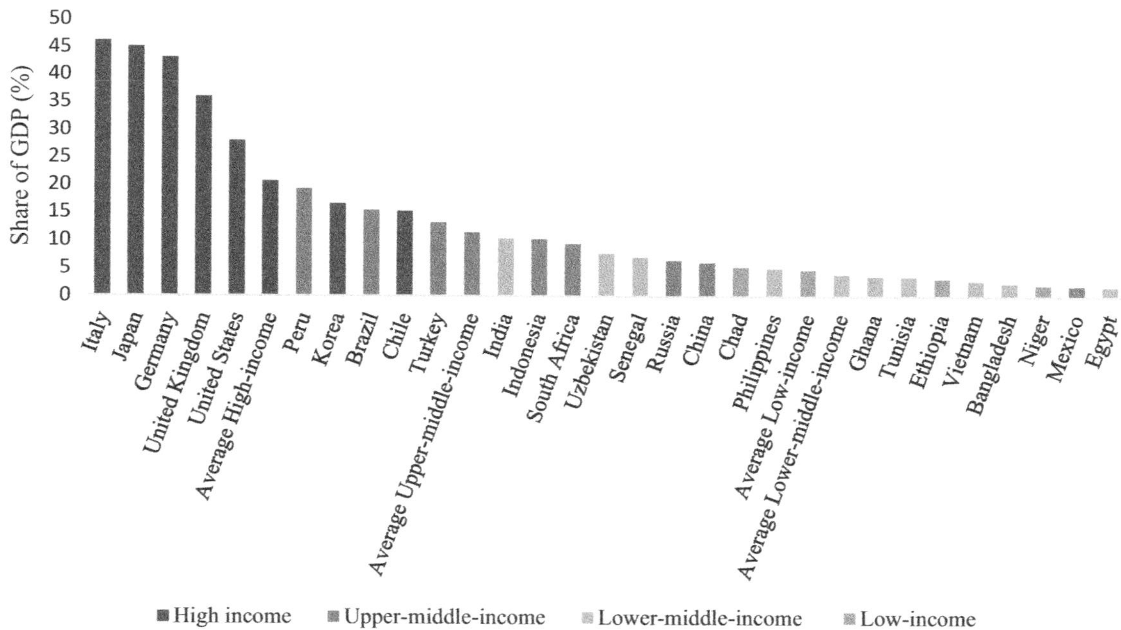

Fig. 2.4 Size of the fiscal response to COVID-19 for country income groups as percent of GDP (*Note* The figure shows the total fiscal support which sums the "above-the-line-measures" in terms of additional spending or forgone revenues and the subtotal of liquidity support measures, both represented as a percent of GDP. *Source* Authors based in the International Monetary Fund, Fiscal monitor database of country: Fiscal measures in response to the COVID-19 pandemic, October 2021)

thus necessitates proactive management, with any delay resulting in longer-term economic and social consequences that are similar to previous debt crises such as the declining performance of the economy, stagnation, and the diversion of essential public spending on education, healthcare, and social safety nets. This, in turn limits the potential role of financing in stimulating growth, and risks opportunities of improved debt sustainability in the long run (Mohieldin, February 2022; United Nations, 2022; World Bank, 2022).

COVID-19 Stimulus Packages and Climate Finance

Massive financial emergency and recovery packages were introduced during COVID-19 at an exceptional speed. These packages were expected to have long-term impacts on the economy, the environment, and the people. Given the effect of the pandemic on most economies, and the urgent, simultaneous risk of the climate crisis, these stimulus packages represented an opportunity for an inclusive, green, and resilient recovery. This particularly applies if they were well-designed, climate-friendly, and green (Agrawala et al., 2020).

The European Union stimulus package worth EUR 2.018 trillion, consists of a long-term budget for 2021–2027 and a NextGenerationEU temporary instrument. The package devotes 30% to fighting climate change, and fair climate transition and preparedness, with over 50% of the stimulus allocated toward recovery and resilience (European Commission, April 2021). The United Kingdom's "Ten Point Plan for a Green Industrial Revolution" aims to mobilize GBP 12 billion of government investment to create and support up to 250,000 highly skilled green jobs, which may potentially encourage over three times as much private sector investment by 2030 (HM Government, November 2020). Similarly, at the heart of the South Korea's "New Deal" stimulus is Korea's "Green New Deal," which targets strengthening climate action and realizing a green economy through investments that focus on green infrastructures, renewable energy, and growing green industries (Government of the Republic of Korea, July 2020)

However, a comprehensive analysis of COVID-19-related fiscal recovery packages shows a rather different story. Analysis of fiscal recovery efforts of 50 leading economies, worth of USD 14.6 trillion in 2020, reveals that only USD 368 billion is green[2]. Moreover, only 18% of recovery spending is expected to contribute to reducing GHG emissions, representing 2.5% of the announced total spending. Furthermore, despite the importance of research and development (R&D) in meeting the Paris Agreement targets, only USD 28.9 billion was directed toward green R&D (O'Callaghan & Murdock, 2021). A wider analysis of the USD 17.2 trillion in stimulus as of July 2021 shows that it will have a net negative environmental impact in 15 of the G20 countries, and in half of the other studied countries. In many cases, particularly in emerging markets, such stimulus does not have any green orientation and only further contributes to high-carbon and low resilience activities. Also, nature and biodiversity loss have been side-lined in most of the announced stimulus (Vivid Economics, July 2021). Overall, most governments missed the opportunity to use the COVID-19 stimulus to put their economies on the necessary climate action trajectory.

It is also evident that the USD 100 billion climate finance pledge remains largely unmet. Adopting the most generous interpretation of climate finance, the latest report from the OECD reveals that the total climate finance reached only USD 79.9 billion in 2019, which was "disappointing" (Mitchell, November 2021; OECD, September 2021). In contrast to these numbers, the analysis undertaken by the non-governmental organization Oxfam estimated that the total climate finance for 2017–2018 was around USD 60 billion below the OECD's 2018 estimate, suggesting that developed countries will miss the USD 100 billion yearly targets even up to 2025—the newly announced deadline for the pledge (Oxfam, 2020, 2021). The gap in the numbers of climate financing highlights the definitional and methodological issues that plague the matter, as there is a no clear universal definition of what should count as climate finance (Chowdhury & Jomo, 2022). This suggests that it is time to have unified international standards that can be utilized to follow country pledges (Mohieldin, November 2021). This comes in light of the move by the United Nations Secretary-General to launch an Expert Group to develop clear standards for net-zero emission pledges by non-state entities and hasten their implementation (United Nations, March 2022). It is also important to note that the USD 100 billion pledge was somewhat arbitrary to begin with and does not meet the full needs of developing nations. For instance, USD 1.6–3.8 trillion alone are needed in annual energy investments to meet the 1.5°C goal of the Paris Agreement, and countries are spending less than half of this requirement (Alayza & Caldwell, 2021). Additionally, there has been no guarantee for climate finance modality to be provided in the next years by wealthy countries despite emphasis on the need for concessional funding. This accompanies statistics that show loans constituted 71% of public finance in 2019. Overdependence on loans increases Low- and Middle-Income Countries' (LMICs) reliance on debt, and exaggerates the true value of climate finance mobilized (Chowdhury & Jomo, 2022). Moreover, this pattern risks pushing these countries into 'climate debt traps,' where facing climate-related disasters will be accompanied with surges in national borrowing at a potentially higher cost for LMICs (Chowdhury & Jomo, 2022; Picolotti & Miller, November 2020).

Aligning Public Finance with the SDGs

Public finance can affect the SDGs by providing incentives for increased sustainable production or consumption patterns, as well as through generating revenues that can be harnessed to finance the goals while reducing inequalities. Regarding the latter, it is important to note that taxing and spending are intended to be jointly considered in order to provide a holistic assessment of the aggregate effect of fiscal policy changes on addressing inequalities. In this matter, governments can make use of the microdata available to study the impact of taxes and transfers, while feeding their findings into designing fiscal frameworks that reduce poverty and inequality. The United Nations Statistical Commission adopted a methodology for evaluating the redistributive impact of fiscal policy by comparing pre- and post-fiscal incomes and measuring the change in inequality by the Gini coefficient (United Nations, 2022). Climate change mitigation and adaptation can be incentivized by fiscal tools and regulatory policies. The tools that policymakers can use include price mechanisms such as taxes, cap-and-trade systems, and subsidies; regulations such as energy efficiency standards; and public investment and guarantees (United Nations, 2022). For example, as of 2020, there were approximately 229 biodiversity-relevant taxes, of which 206 are currently in effect (Dasgupta, 2021, p. 469). However, it is evident that governments still support activities that hinder progress to the SDGs. For example, in many countries most of the public finance directed toward subsidies goes into activities that are detrimental to ecosystems and biodiversity (Dasgupta, 2021, pp. 468–470). Recent estimates have shown that the true environmental costs of some subsidies largely outweigh their monetary value. When accounting for the negative externalities arising from fossil fuel subsidies, the resulting aggregate cost of such subsidies is estimated to be around USD 5.2 trillion annually (6.5% of GDP in 2017) (Coady et al., 2019). Consequently, the shift toward more green fiscal policies would need careful design and sequencing to ensure a just transition. Countries will need to introduce supporting policies to support livelihoods, ensure energy affordability, and expand social protection schemes to alleviate increases in poverty that might result from climate change mitigation policies.

Governments have yet to align their fiscal systems with the SDGs and climate action (United Nations, 2022). The most crucial element of this alignment is for the budget to complement or ideally base itself on the SDGs. Public budgets serve as economical, legal, political, and accountability tools that guarantee the continuity of policy objectives beyond electoral cycles while balancing both short and long-term needs. They represent the most powerful instruments to ensure that policies and plans are effectively implemented. Accordingly, budgeting for the SDGs entails explicit and measurable presentation of SDG targets in the budgetary framework throughout its different stages of strategic planning, budget formulation, budget execution, and monitoring and reviewing. Simply put, if governments are committed to delivering the SDGs, it is insufficient to simply announce this commitment in high-level policy papers. They must also accompany this with action to bridge the SDGs through national budget systems, as this is 'the strongest domestic legal basis that reflects countries' commitments to 2030 Agenda' (OECD, 2019c; UNDP, 2020b). Otherwise, there may be a disconnection between the strategic planning frameworks that have made commitments to the 2030 Agenda, and budgets that would follow the 'business as usual' track (UNDP, 2020b). Budgeting for the SDGs also provides a mechanism for integrating the country's long-term sustainable development aspirations and the resources controlled by the government. As such, incorporating the SDGs within the budget is a key objective of the INFF and can have important implications on other aspects and flows by promoting deeper alignment and resources from other actors, especially the private sector. Despite the well-established consensus on the importance of the budget-SDG integration, there remains some discrepancy between the expressed political support for the SDGs and having the goals reflected in national budgets. Furthermore, G20 countries appear to be lagging behind in linking their budgets to the SDGs, a step which is particularly important to this group as they constitute two-thirds of the world's population and 85% of the global GDP (Sachs et al., 2022).

A survey of the current approaches to SDG budgeting shows that such approaches remain highly fragmented, with most focused on thematic areas such as climate change or gender-responsive policies. As such, most SDG and climate budgeting tools do not help prioritize budget allocations and expenditures across different sectors. Methodological challenges related to how to account for synergies and trade-offs among different policies and expenditure proposals still persist in most current approaches. Moreover, aspects of the budget cycle are receiving more focus than others, as seen with procedures such as tagging in comparison to the limited guidance around integrating climate objectives and the SDGs in the budget call circulars. This limited guidance hinders practical implementation of SDG-budget integration as the budget call circular is the most important guidance document; it has operational guidelines, targets and provides a clear set of rules for the budget process (Gonguet et al., 2021). Additionally, the output from the monitoring and review stage of the budget do not feed into the subsequent planning and budgeting activities. The in-year budget execution phase of the budget cycle remains largely understudied despite how the former can be the major cause of poor program and project implementation as well as result in inefficient utilization of resources and failure to translate strategic priorities to results (World Bank, 1998). It is recommended that the move toward SDG budgeting should be integrated with an existing Public Financial Management (PFM) agenda or in the broader context of PFM reform. It should also be supported by a legal foundation that outlines objectives with a clear mandate for the Ministry of Finance and line ministries that lays out key requirements of the move (Gonguet et al., 2021; UN DESA, 2019). Countries such as Mexico and New Zealand have introduced amendments to their PFM legal frameworks to incorporate the priority of SDGs and NDCs. On a more limited scale, states like Sweden have PFM elements in their climate laws while Albania, Canada, Morocco, and Uganda have these elements in gender-related laws (Huang et al., 2020; Rivero del Paso, January 2021; UN Women, 2021).

Concluding Remarks

Advancing progress toward the 2030 Agenda calls for actions by national governments that promote a whole-of-government approach and foster the engagement of a wide range of stakeholders. While the pandemic has been a major setback to the SDGs, it has exposed systematic vulnerabilities and failures that are addressed at the heart of the 2030 Agenda. The pandemic thus presents a chance for governments to "Build Forward Better" by emphasizing what is necessary for recovery: inclusion, equity, growth, resilience, and sustainability. Within this framework, national governments must enhance policy coherence by realizing the synergies and trade-offs across different sectors and goals. Implementation at the last mile calls for national governments to establish clear coordination with LRGs and enable them by facilitating the provision of the needed funds. Furthermore, attracting the private sector to sustainability initiatives and harnessing the sector's productivity and financial capacity can be achieved through government action. Specifically, through state efforts of unifying and mandating corporate sustainability disclosure requirements and aligning fiscal tools with sustainability and climate action.

Timely, disaggregated, and comparable data is a prerequisite for designing effective policies and reaching those who are farthest behind. Data allows for the measuring of policy and program impacts while highlighting the spillovers and thus contributing to the needed policy coherence. However, large inequalities exist across countries in their data and statistical capacities and there is a gap between the expressed importance of data and the investments directed into data systems. The cross-cutting theme across this chapter is that adequate finance is a major key toward realizing the 2030 Agenda's ambitions. The need for finance goes beyond mere funding to the need for effective institutions and well-coordinated public policies that balance the short-run immediate needs with the long-run investments in infrastructure, human capital, and resilience. Countries have yet to realize the benefit of SDG budgeting despite the acknowledgment that budgets represent a milestone in financing the progress toward the goals and mobilizing the necessary finance from relevant actors. This is key given that the ambitions of the 2030 Agenda put pressure on public sector resources, making the role of the private sector indispensable. This role has two aspects, good governance in business practices and unlocking private investment in sustainable development.

Acknowledgements The authors would like to acknowledge the efforts of Haajar Abu Ismail and her editorial work in this chapter.

References

Agrawala, S., Dussaux, D., & Monti, N. (2020). *What policies for greening the crisis response and economic recovery? Lessons learned from past green stimulus measures and implications for the COVID-19 crisis* (OECD Environment Working Papers No. 164). OECD Publishing. https://doi.org/10.1787/c50f186f-en

Alayza, N., & Caldwell, M. (2021). *Financing climate action and the COVID-19 pandemic: An analysis of 17 developing countries* (Working Paper). World Resources Institute. https://files.wri.org/d8/s3fs-public/2021-10/financing-climate-action-covid-19-pandemic.pdf?VersionId=TMOS7TyYBnE9FAhjVZWubqBGrc_k1ycm

Allen, C., Metternicht, G., & Wiedmann, T. (2018). Initial progress in implementing the Sustainable Development Goals (SDGs): A review of evidence from countries. *Sustainability Science, 13*, 1453–1467. https://doi.org/10.1007/s11625-018-0572-3

Basdevant, O., Chaponda, T., Gonguet, F., Honda, J., & Thomas, S. (2020). Designing fiscal rules to protect public investment. In G. Schwartz, M. Fouad, T. Hansen, & G. Verdier (Eds.), *Well spent: How strong infrastructure governance can end waste in public investment?* International Monetary Fund.

Bauer, R., Ruof, T., & Smeets, P. (2021). Get real! Individuals prefer more sustainable investments. *Review of Financial Studies, 34*(8), 3976–4043. https://doi.org/10.1093/rfs/hhab037

Baum, A., Mogues, T., & Verdier, G. (2020). Getting the most from public investment. In G. Schwartz, M. Fouad, T. Hansen, & G. Verdier (Eds.), *Well spent: How strong infrastructure governance can end waste in public investment?* International Monetary Fund.

Bettelli, P. (2021, January). *What the world learned setting development Goals*. International Institute for Sustainable Development. Retrieved May 30, 2022, from https://www.iisd.org/articles/deep-dive/what-world-learned-setting-development-goals

Bouyé, M., Harmeling, S., & Schulz, N. (2018). *Connecting the dots: Elements for a joined-up implementation of the 2030 agenda and Paris agreement*. Deutsche Gesellschaft fur Internationale Zusammenarbeit and World Resources Institute.

Chen, R. (2021). *A demand-side view of mobile internet adoption in the global south* (Policy Research Working Paper 9590). World Bank. http://documents.worldbank.org/curated/en/492871616350929155/A-Demand-Side-View-of-Mobile-Internet-Adoption-in-the-Global-South

Chowdhury, A., & Jomo, K. S. (2022). The climate finance conundrum. *Development, 65*, 29–41. https://doi.org/10.1057/s41301-022-00329-0

Coady, D., Parry, I., Le, N.-P., & Shang, B. (2019). *Global fossil fuel subsidies remain large. An update based on country-level estimates* (IMF Working Paper 19/89). International Monetary Fund.

Dahmm, H. (2021, April 21). *Major environmental data gaps remain, but progress is on the horizon.* The Sustainable Development Solution Network's Thematic Research Network on Data and Statistics (SDSN TReNDS). https://www.sdsntrends.org/blog/2021/environmentaldatagaps?locale=en

Dang, H.-A. H., Pullinger, J., Serajuddin, U., & Stacy, B. (2021). *Statistical performance indicators and index: A new tool to measure country statistical capacity.* World Bank. https://openknowledge.worldbank.org/handle/10986/35301

Dargent, E., Lotta, G., MejíaGuerra, J. A., & Moncada, G. (2020). *Who wants to know? The political economy of statistical capacity in Latin America.* Inter-American Development Bank.

Dasgupta, P. (2021). *The economics of biodiversity: The Dasgupta review.* HM Treasury.

de Mello, L. (2016). *Co*-ordination and implementation of the Sustainable Development Goals: The role of the centres of government. In P. Love (Ed.), *Debate the issues: New Approaches to economic challenges.* OECD Publishing.

European Commission. (2018). *Delivering the Sustainable Development Goals at local and regional level.* Recommendations to the European Commission by the subgroup on "SDGs at local and regional level" of the Multi-Stakeholder Platform on the Implementation of the Sustainable Development Goals in the EU. https://ec.europa.eu/info/sites/default/files/delivering-sdgs-local-regional-level.pdf

European Commission. (2021, April). *The EU's 2021–2027 long-term budget and NextGenerationEU: Facts and figures.* European Commission.

Flyvbjerg, B. (2009). Survival of the unfittest: Why the worst infrastructure gets built—And what we can do about it. *Oxford Review of Economic Policy, 25*(3), 344–367. https://doi.org/10.1093/oxrep/grp024

Gasper, V., Amaglobeli, D., Garcia-Escribano, M., Prady, D., & Soto, M. (2019). *Fiscal policy and development: Human, social, and physical investments for the SDGs* (IMF Staff Discussion Notes No. 19/03). International Monetary Fund.

Gentilini, U., Almenfi, M., Orton, I., & Dale, P. (2020). *Social protection and jobs responses to COVID-19: A real-time review of country measures.* World Bank, Washington, DC. https://openknowledge.worldbank.org/handle/10986/33635

Gonguet, F., Wendling, C. P., Sakrak, O. A., & Battersby, B. (2021). *Climate-sensitive management of public finances—"Green PFM"* (IMF Staff Climate Note No 2021/002). International Monetary Fund.

Gornitzka, C. P., & Pipa, A. F. (2018, April 24). *A new type of leadership from national governments is essential for success of the SDGs*. Brookings. Retrieved May 30, 2022, from https://www.brookings.edu/blog/up-front/2018/04/24/a-new-type-of-leadership-from-national-governments-is-essential-for-success-of-the-sdgs/

Government of the Republic of Korea. (2020, July). *The Korean new deal national strategy for a great transformation*. https://english.moef.go.kr/pc/selectTbPressCenterDtl.do?boardCd=N0001s&seq=4948

Guo, W., & Quayyum, S. (2020). Building resilience to natural disaster in vulnerable states: Savings from ex-ante interventions. In G. Schwartz, M. Fouad, T. Hansen, & G. Verdier (Eds.), *Well spent: How strong infrastructure governance can end waste in public investment?* International Monetary Fund.

Hammer, C., Morales, L. G., Contreras-Gonzalez, I. M. & Wollburg, P. R. (2021, August). National statistical offices still face disruptions and challenges as they adapt to a "new normal". *World Bank Blogs*. https://blogs.worldbank.org/opendata/national-statistical-offices-still-face-disruptions-and-challenges-they-adapt-new-normal

HM Government. (2020, November). *The ten point plan for a green industrial revolution: Building back better, supporting green jobs, and accelerating our path to net zero*. https://assets.publishing.service.gov.uk/government/uploads/system/uploads/attachment_data/file/936567/10_POINT_PLAN_BOOKLET.pdf

Huang, C., de Renzio, P., & McCullough, D. (2020). *New Zealand's "well-being budget": A new model for managing public finances?* International Budget Partnership. https://www.internationalbudget.org/wp-content/uploads/new-zealand-well-being-budget-may-v2-2020.pdf

International Monetary Fund (IMF). (2019). *World Economic Outlook, October 2019: Global manufacturing downturn, rising trade barriers*. International Monetary Fund.

International Telecommunication Union (ITU), World Bank, Commonwealth Secretariat (ComSec), Commonwealth Telecommunications Organisation (CTO), & Cooperative Cyber Defence Centre of Excellence, North Atlantic Treaty Organization (NATO). (2018). *Guide to developing a national cybersecurity strategy: Strategic engagement in cybersecurity*. ITU. https://www.itu.int/dms_pub/itu-d/opb/str/D-STR-CYB_GUIDE.01-2018-PDF-E.pdf

Kitzmueller, L., Stacy, B., & Mahler, D. G. (2021, August 10). *Are we there yet? Many countries don't report progress on all SDGs according to the World Bank's new Statistical Performance Indicators*. World Bank Blogs. https://blogs.worldbank.org/opendata/are-we-there-yet-many-countries-dont-report-progress-all-sdgs-according-world-banks-new#:~:text=No%20country%20reported%20data%20on,for%202019%20than%20for%202015

Martínez-Solimán, M., & Mohieldin, M. (2016). Lessons and recommendations from the CEB reviews: A forward-looking perspective from the MDG Era. In *Transitioning from the MDGs to the SDGs*. United Nations Development Programme and the World Bank Group.

Mitchell, I. (2021, November 5). *Climate finance is the elephant in the room at COP26*. Chatham House blog. Retrieved 29 May 2022 from https://www.chathamhouse.org/2021/11/climate-finance-elephant-room-cop26

Mohieldin, M. (2017, March 20). Tackling inequality: What can governments do to generate growth that benefits everyone equally? In *SDGs from promise to practice: A UNA-UK publication providing analysis and recommendations on achieving the Sustainable Development Goals*. Retrieved 29 May 2022 from http://www.sustainablegoals.org.uk/tackling-inequality/

Mohieldin, M. (2021, March 15). IMF executive director letter: UN goals offer best route to combat climate crisis. *Financial Times*. Retrieved May 29, 2022, from https://www.ft.com/content/5faccf6c-01f2-4295-8a3c-ba7940e26ff6

Mohieldin, M. (2021, April 21). *Climate change and rigid policies*. *Ahram Online*. Retrieved May 29, 2022, from https://english.ahram.org.eg/NewsContent/4/0/407642/Opinion/0/Climate-change-and-rigid-policies.aspx

Mohieldin, M. (2021, May 17). *A global economic system in turmoil: What is at stake for the MENA region?* [Conference presentation]. Economic Research Forum 27th Annual Conference. Retrieved May 29, 2022 from https://www.researchgate.net/publication/351634820_A_Global_Economic_System_in_Turmoil_What_is_at_Stake_for_the_MENA_region

Mohieldin, M. (2021, November 28). COP26 achieves less than hoped. *Ahram Online*. Retrieved May 29, 2022, from https://english.ahram.org.eg/NewsContentP/50/440767/AlAhram-Weekly/COP-achieves-less-than-hoped.aspx

Mohieldin, M. (2022, February 28). Financing a new world order. *Ahram Online*. Retrieved May 29, 2022, from https://english.ahram.org.eg/NewsContentP/50/461580/AlAhram-Weekly/Financing-a-new-world-order.aspx

Mohieldin, M., & Shehata, M. (2021). The SDGs as an operational framework for post COVID-19 response and recovery. *AD-Minister, 38*, 5–42. https://doi.org/10.17230/Ad-minister.38.1

Morita, K., Okitasari, M., & Masuda, H. (2020). Analysis of national and local governance systems to achieve the sustainable development goals: Case studies of Japan and Indonesia. *Sustainability Science, 15*, 179–202. https://doi.org/10.1007/s11625-019-00739-z

O'Callaghan, B. J., & Murdock, E. (2021). *Are we building back better? Evidence from 2020 and pathways to inclusive green recovery spending*. United Nations Environment Programme and Oxford's Economic Recovery Project. https://wedocs.unep.org/bitstream/handle/20.500.11822/35281/AWBBB.pdf

OECD. (2008). *Conducting sustainability assessments*. OECD Sustainable Development Studies. OECD Publishing. https://doi.org/10.1787/9789264047266-en

OECD. (2017, July 18). *Getting governments organised to deliver on the Sustainable Development Goals. Summary reports and key steps*. High Level Political Forum. United Nations.

OECD. (2019a). *Enhancing access to and sharing of data: Reconciling risks and benefits for data re-use across societies*. OECD Publishing. https://doi.org/10.1787/276aaca8-en

OECD. (2019b). *The path to becoming a data-driven public sector, OECD digital government studies*. OECD Publishing. https://doi.org/10.1787/059814a7-en

OECD. (2019c). *Governance as an SDG accelerator: Country experiences and tools*. OECD Publishing. https://doi.org/10.1787/0666b085-en

OECD. (2020a). *Building a coherent response for a sustainable post-COVID-19 recovery,* OECD Policy Responses to Coronavirus (COVID-19). Retrieved May 29, 2022 from https://read.oecd-ilibrary.org/view/?ref=376_376369-eayqu00bgf&title=Building-a-coherent-response-for-a-sustainable-post-COVID-19-recovery

OECD. (2020b). *The territorial impact of COVID-19: Managing the crisis across levels of government*. OECD Policy Responses to Coronavirus (COVID-19). Retrieved May 29, 2022 from https://www.oecd.org/coronavirus/policy-responses/the-territorial-impact-of-covid-19-managing-the-crisis-across-levels-of-government-d3e314e1/

OECD. (2020c). *The impact of the COVID-19 crisis on regional and local governments: Main findings from the joint CoR-OECD survey* (OECD Regional Development Papers, No. 05). OECD Publishing. https://doi.org/10.1787/fb952497-en

OECD. (2020d). *Strategic foresight for the COVID-19 crisis and beyond: Using futures thinking to design better public policies*. OECD Policy Responses to Coronavirus (COVID-19). Retrieved May 29, 2022 from https://www.oecd.org/coronavirus/policy-responses/strategic-foresight-for-the-covid-19-crisis-and-beyond-using-futures-thinking-to-design-better-public-policies-c3448fa5/

OECD. (2020e). *Global outlook on financing for sustainable development 2021: A new way to invest for people and planet*. OECD Publishing. https://doi.org/10.1787/e3c30a9a-en

OECD. (2021, September 17). *Statement from OECD secretary-general Mathias Cormann on climate finance in 2019* (Press Release). Retrieved 29 May 2022 from https://www.oecd.org/newsroom/statement-from-oecd-secretary-general-mathias-cormann-on-climate-finance-in-2019.htm

OECD. (2021). *Data for development profiles: Official development assistance for data and statistical systems*. OECD Publishing. https://doi.org/10.1787/84baa8f3-en

Okitasari, M., Sunam, R., Mishra, R., Masuda, H., Morita, K., & Takemoto, K. N. (2019). Governance and national implementation of the 2030 agenda: Lessons from voluntary national reviews. *United Nations University Institute for the Advanced Study of Sustainability, 18*, 1–4.

Open Data Watch (ODIN). (2021). *ODIN Open Data Inventory 2020/21*. https://odin.opendatawatch.com/Report/annualReport2020#sec1

Oxfam. (2020). *Climate finance shadow report 2020: Assessing progress towards the $100 billion commitment*. Oxfam. https://oxfamilibrary.openrepository.com/bitstream/handle/10546/621066/bp-climate-finance-shadow-report-2020-201020-en.pdf

Oxfam. (2021). *Poorer nations expected to face up to $75 billion six-year shortfall in climate finance: Oxfam* (Press Release). Retrieved 30 May 2022 from https://www.oxfam.org/en/press-releases/poorer-nations-expected-face-75-billion-six-year-shortfall-climate-finance-oxfam

Partnership in Statistics for Development in the 21st Century (PARIS21). (2018). *Survey results: New approaches to capacity development and future priorities, CD4.0 Survey*. PARIS21. https://paris21.org/capacity-development-40/cd40-survey

Partnership in Statistics for Development in the 21st Century (PARIS21). (2021). *The partner report on support to statistics 2021*. Paris.

Picolotti, R., & Miller, A. (2020, November 5). *Debt-for-climate swaps can help developing countries make a green recovery*. Climate Home News. Retrieved 29 May 2022 from https://www.climatechangenews.com/2020/11/05/debt-climate-swaps-can-help-developing-countries-make-green-recovery/

Rajaram, A., Kai, K., Tuan, M. L., Jay-Hyung, K., & Jonas, F. (2014). *The power of public investment management: Transforming resources into assets for growth*. World Bank.

Rivero del Paso, L. (2021, January 28). *Leveraging public budgets for the Sustainable Development Goals: The case of Mexico*. IMF PFM Blog. Retrieved 29 May 2022 from https://blog-pfm.imf.org/pfmblog/2021/01/-leveraging-public-budgets-for-the-sustainable-development-goals-the-case-of-mexico-.html

Sachs, J., Lafortune, G., Kroll, C., Fuller, G., & Woelm, F. (2022). *Sustainable Development Report 2022: From crisis to sustainable development: The SDGs as roadmap to 2030 and beyond*. Cambridge University Press.

Sachs, J., Schmidt-Traub, G., Kroll, C., Lafortune, G., & Fuller, G. (2021). *Sustainable Development Report 2021: The decade of action for the Sustainable Development Goals*. Cambridge University Press.

Schwartz, G., Fouad, M., Hansen, T., & Verdier, G. (2020). Well spent: How strong infrastructure governance can end waste in public investment? In G.

Schwartz, M. Fouad, T. Hansen, & G. Verdier (Eds.), *Well spent: How strong infrastructure governance can end waste in public investment?* International Monetary Fund.

Shawoo, Z., Dzebo, A., Hägele, R., Iacobuta, G., Chan, S., Muhoza, C., Osano, P., Francisco, M., Persson, Å., Linner, B., & Vijge, M. J. (2020). *Increasing policy coherence between NDCs and SDGs: A national perspective*. Stockholm Environment Institute.

Stiglitz, J., Fitoussi, J., & Durand, M. (2018). *Beyond GDP: Measuring what counts for economic and social performance*. OECD Publishing. https://doi.org/10.1787/9789264307292-en

Stiglitz, J., Sen, A., & Fitoussi, J. (2009). *Report by the commission on the measurement of economic performance and social progress.*

Sustainable Development Solutions Network (SDSN). (2016). *Getting started with the SDGs in cities: A guide for stakeholders*. https://resources.unsdsn.org/getting-started-with-the-sdgs-in-cities

Task Force on Climate-related Financial Disclosures (TCFD). (2021). *2021 status report*. TCFD.

The Economist. (2022, February 12). Who buys the dirty energy assets public companies no longer want? *The Economist*. Retrieved May 30 2022 from https://www.economist.com/finance-and-economics/who-buys-the-dirty-energy-assets-public-companies-no-longer-want/21807594

The Sustainable Development Solution Network's Thematic Research Network on Data and Statistics (SDSN TReNDS). (2017). *Counting on the world: Building modern data system for sustainable development*. SDSN TReNDS.

United Cities and Local Governments (UCLG). (2017). *National and sub-national governments on the way towards the localization of the SDGs*. Report Towards the 2017 HLPF. Global Task Force of Local and Regional Governments. https://www.uclg.org/sites/default/files/localgov_report_localizationsdg_hlpf.pdf

United Cities and Local Governments (UCLG). (2019). *Towards the localization of the SDGs. Local and regional governments.* Report Towards the 2019 HLPF. 3rd Report. Global Task Force of Local and Regional Governments. https://www.uclg.org/sites/default/files/towards_the_localization_of_the_sdgs_0.pdf

United Nations. (2015a). *Transforming our world: The 2030 Agenda for Sustainable Development. Resolution adopted by the General Assembly on 25 September 2015*. A/RES/70/1. U.N. General Assembly, Seventieth Session. Agenda items 15 and 116. United Nations.

United Nations. (2015b). *Addis Ababa Action Agenda of the third international conference on financing for development.* United Nations.

United Nations. (2020a). *The Sustainable Development Goals Report 2020.* United Nations.

United Nations. (2020b). *Report of the intersecretariat working group on national accounts* (E/CN.3/2021/8). United Nations.

United Nations. (2021a). *Inter-agency task force on financing for development, financing for Sustainable Development Report 2021.* United Nations. https://developmentfinance.un.org/fsdr2021

United Nations. (2021b). *The Sustainable Development Goals Report 2021.* United Nations.

United Nations. (2022, March 21). *Expert group on the net-zero emissions commitments of non-state entities* (Press Release). Retrieved 20 May 2022 from https://www.un.org/sg/en/content/sg/personnel-appointments/2022-03-31/expert-group-the-net-zero-emissions-commitments-of-non-state-entities%C2%A0

United Nations. (2022). *Inter-agency task force on financing for development financing for Sustainable Development Report 2022.* United Nations. https://developmentfinance.un.org/fsdr2022

United Nations Conference on Trade and Development (UNCTAD). (2014). *World Investment Report 2014: Investing in the SDGs: An action plan.* United Nations.

United Nations Conference on Trade and Development (UNCTAD). (2022). *Data protection and privacy legislation worldwide* (Dashboard). Retrieved 29 May 2022 from https://unctad.org/page/data-protection-and-privacy-legislation-worldwide

United Nations Department of Economic and Social Affairs (UN DESA). (2019). *World public sector report 2019: Sustainable Development Goal 16: Focus on public institutions.* United Nations. https://www.un-ilibrary.org/content/books/9789210041409#chapters

United Nations Department of Economic and Social Affairs (UN DESA). (2021a). *World Public Sector Report 2021: National institutional arrangements for implementation of the Sustainable Development Goals: A five-year stocktaking.* United Nations. https://www.un.org/en/desa/world-public-sector-report

United Nations Department of Economic and Social Affairs (UN DESA). (2021b). *Horizontal and vertical integration are more necessary than ever for COVID-19 recovery and SDG implementation* (Policy Brief No. 115). https://www.un.org/development/desa/dpad/publication/un-desa-policy-brief-115-horizontal-and-vertical-integration-are-more-necessary-than-ever-for-covid-19-recovery-and-sdg-implementation/

United Nations Development Programme (UNDP). (2020a). *COVID-19 and human development: Assessing the crisis, envisioning the recovery.* United Nations.

United Nations Development Programme (UNDP). (2020b). *Budgeting for the Sustainable Development Goals: Aligning domestic budgets with the SDGs: A*

Guidebook. https://sdgfinance.undp.org/sites/default/files/UNDP%20Budgeting%20for%20the%20SDGs%20-%20Guidebook_Nov%202020.pdf

United Nations Development Programme (UNDP) & United Nations Capital Development Fund (UNCDF). (2016). *Getting to the last mile in least developed countries*United Nations Development Programme (UNDP) and United Nations Capital Development Fund (UNCDF).

United Nations Statistics Division (UNSD). (2021). *IAEG-SDGs: Tier classification for global SDG indicators*. Updated 29 March 2021 by the Inter-agency and Expert Group on SDG Indicators. https://unstats.un.org/sdgs/files/Tier%20Classification%20of%20SDG%20Indicators_29%20Mar%202021_web.pdf

United Nations Sustainable Development Group (UNSDG) (2020). *Education during COVID-19 and beyond* (Policy Brief). https://unsdg.un.org/sites/default/files/2020-08/sg_policy_brief_covid-19_and_education_august_2020.pdf

United States Department of Labor. (2021, October 13). *US Department of Labor proposes rule to remove barriers to considering environmental, social, governance factors in plan management* (Press Release). Retrieved 29 May 2022 from https://www.dol.gov/newsroom/releases/ebsa/ebsa20211013

UN Women. (2021). *COVID-19 and fiscal policy: Applying gender-responsive budgeting in support and recovery* (Policy Brief No. 2021).

Vivid Economics. (July 2021). *Greenness of Stimulus Index: An assessment of COVID-19 stimulus by G20 countries and other major economies in relation to climate action and biodiversity goal*. https://www.vivideconomics.com/casestudy/greenness-for-stimulus-index/

Wanckel, C. (2021). Introducing a digital tool for sustainability impact assessments within the German Federal Government: A neo-institutional perspective. *International Review of Administrative Sciences*. https://doi.org/10.1177/00208523211047093

World Bank. (1998). *Public expenditure management handbook*. The World Bank Group.

World Bank. (2018, August 3). *Investing in people to build human capital*. Retrieved 29 May 2022 from https://www.worldbank.org/en/news/immersive-story/2018/08/03/investing-in-people-to-build-human-capital

World Bank. (2021a). *Investing in human capital for a sustainable recovery: The role of public finance*. The World Bank Group.

World Bank. (2021b). *World Development Report 2021: Data for better lives*. The World Bank Group.

World Bank. (2021, October 29). *World Bank: Pandemic threatens to drive unprecedented number of children into learning poverty* (Press Release).

Retrieved 29 May 2022 from https://www.worldbank.org/en/news/press-release/2021/10/29/world-bank-pandemic-threatens-to-drive-unprecedented-number-of-children-into-learning-poverty

World Bank. (2022). *World Development Report 2022: Finance for an equitable recovery*. The World Bank Group.

World Bank & United Nations. (2017). *Combatting cybercrime: Tools and capacity building for emerging economies*. World Bank. https://openknowledge.worldbank.org/handle/10986/30306

World Bank, UNSD & Paris21. (2022). *Survey on the implementation of the Cape Town global action plan for sustainable development data* (English). World Bank Group. http://documents.worldbank.org/curated/en/826351643712794722/Survey-on-the-Implementation-of-the-Cape-Town-Global-Action-Plan-for-Sustainable-Development-Data

CHAPTER 3

How Businesses Can Accelerate and Scale-Up SDG Implementation by Incorporating ESG into Their Strategies

Abstract Businesses are confronted with both global and local risks. Firms' reactions to concerns about sustainable development can minimize risks, boost positive impacts, and benefit society, markets, and ecosystems. Even though the SDGs were designed for governments and not for the private sector, businesses' greatest contribution to the 2030 Agenda is incorporating ESG components into their strategy and therefore scaling up and accelerating SDG implementation. The private sector must be more responsive to the long-term development needs of each location where it directly operates, as well as the needs with downstream or upstream value chain implications. While businesses are stepping up to help achieve the SDGs, acceleration is needed. The world is rapidly changing and problems may intensify. As a result of such changes, businesses must rely on innovation and technology to achieve the SDGs. It is also important that firms measure and demonstrate their contribution toward this achievement. Though a company may have the individual resources to meaningfully contribute to society and the environment, a corporation can only sustainably do so in collaboration with other businesses, CSOs, or the government. When the private sector collaborates with national and local governments on long-term growth and value creation, the result will be much more effective. During the first year of COVID-19, with the added influence of the climate crisis, there was an increased need for corporations to form strong public–private partnerships

M. Mohieldin et al., *Business, Government and the SDGs*,
https://doi.org/10.1007/978-3-031-11196-9_3

with national and local governments, as well as engage in private–private alliances with other enterprises and entrepreneurial endeavors. Progress in the private sector is critical to the 2030 Agenda. Businesses not only shape the market, but also the state and society. Governments, civil society, firms, investors, and consumers must jointly collaborate in achieving global and local goals relating to the SDGs.

Keywords SDG · ESG · Corporate sustainability · Impact investment · Private sector · Business · Net-zero · Climate action · Green recovery · Partnerships

Introduction

The fact that private organizations are powerful actors in the global economy is undeniable. Through the presence of an audience, investors, users, or consumers, global corporations wield the triple-edged sword of economic, political, and social influence. For example, across the largest OECD economies, businesses contribute to 72% of the value-added (Manyika et al., 2021). In comparing companies' market capitalization with countries' GDPs, Apple's net worth would be higher than the GDP of both Italy and Canada, and even nine times higher than the GDP of countries like Finland or Colombia. For example, as shown in Fig. 2.1, Facebook is wealthier than over 150 countries (Lishchuck, 2021).

Figures 3.1 and 3.2 show that the number of Netflix subscribers exceeds the number of inhabitants of Brazil, the amount of Amazon users makes for a larger population than Russia, and Spotify, Disney+ , Tencent, and iQIYI have more subscribers than the entire population of Egypt, Germany, or Vietnam.

There are negative sentiments related to this overwhelming power, particularly when states appear to favor the interests of powerful private entities over the welfare of citizens, as was the case during the 2007–2008 crisis (Miller, 2018). This was most clear during COVID-19, when countries debated the importance of life over the economy. On a similar note, the climate problem stems from this conflict between corporate and vested interests, which is manifested in the flagrant denial of science and reason. It seems that 'power brokers will ensure commitment to economic fairness and environmental sustainability as long as growth and free trade

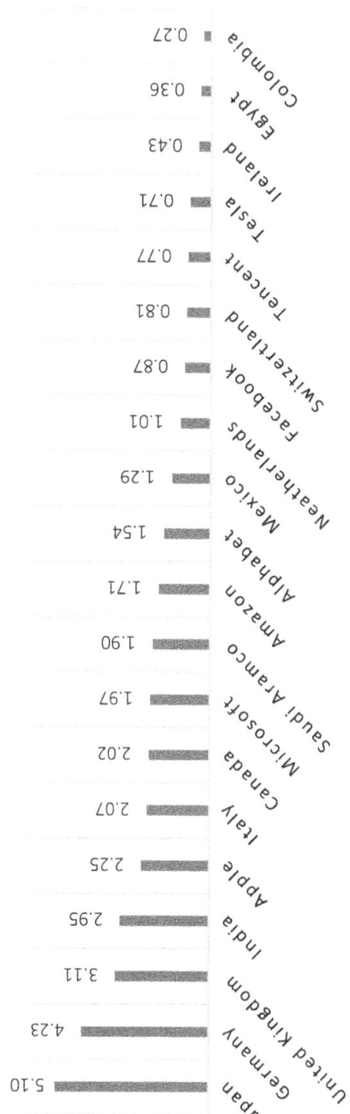

Fig. 3.1 GDP/Market capitalization (in trillion USD) of some countries and firms (*Source* Authors based on different sources)

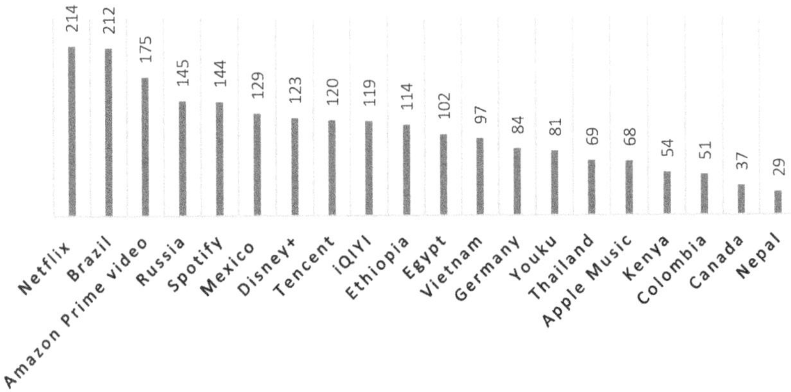

Fig. 3.2 Number of inhabitants/subscribers (millions of people) by December 2021 (*Source* Authors based on various data sources)

are not jeopardized.' People and nature serve corporate expansion and trade efforts when they should be the other way around (Korten, 1998).

The private enterprise remains as both the most important institution in the world and the foundation of Western prosperity (Micklethwait & Wooldridge, 2003). In many circumstances, the power of corporations works to boost efficiency and productivity, influence other firms and people, regulate worldwide production and consumption, and engage in global politics (Hammer, 2001; Micklethwait & Wooldridge, 2003). The argument that all of the former can be done to serve the common good is this chapter's goal.

The Impact of Companies and the Pressure to Internalize Externalities

A company can be defined as an organization, or rather, an organized group of people with common interests—engaged in formal or informal business When formalized, a company is a 'legal entity endowed by the government with certain collective rights and responsibilities' (Micklethwait & Wooldridge, 2003, xvi).

Regardless of their geographical or cyber location, through direct and indirect operations, companies have several impacts: positive and negative, intended and unintended, short and long term. These impacts affect individuals, communities, governments, the natural environment, and markets, which is a matter that these organizations must take responsibility for. In many instances and jurisdictions where firms operate, businesses are held accountable due to mandatory legal regulations. But in several cases, companies, as part of their corporate social responsibility (CSR), have adopted additional actions and voluntary duties in social (human rights, fair labor practices, community relations, etc.), environmental, and economic matters. More recently, both large and small businesses have started to incorporate sustainability in their business strategy (Gonzalez-Perez, 2013).

The production of goods and services and marketing activities generate externalities or impacts (positive and negative) on actors who are not directly involved in the transactions of these activities. These impacts are not necessarily included in market prices nor incorporated into production costs.

In the case of a positive externality, the beneficiaries do not pay for the benefits received. In the case of a negative externality, those affected are not compensated for these damages. However, companies are increasingly required to take responsibility for their operations' environmental and social liabilities as well as to mitigate, repair, regenerate, and provide compensation for their negative externalities to society and nature. To reduce corporate viability risks, businesses internalize their externalities, due to a variety of factors, including current or anticipated regulatory impositions, market demands or opportunities, reputational risk, ESG-driven investment, access to activist consumers' niche markets, stakeholder pressures, and improving relations with governments and local communities.

However, it is often not recognized enough that in production, distribution, trade, and marketing, the private sector generates economic and social value both directly and indirectly in different ways. Shareholders invest capital in a business with the expectation that the activities will add value to the initial investment. Also, companies incur costs in their operations (wages, payments to suppliers, taxes, and financial costs) that directly generate economic value for employees, government taxes, and suppliers of goods, services, and capital, thus rejuvenating economies and society.

The Business Sector's Motivation for Increased Adaptation of ESG Indicators

By the 2020s, societies, investors, and, in some cases, regulators have increased expectations for companies to commit to improving the sustainable development of the communities they operate in. This also applies to how organizations should respond to negative indirect impacts along their value chain if they are unable to root their core business in sustainable practices (Silva et al., 2019). These expectations are even more prominent when associated with climate action, specifically in the case of GHG emissions.

Threats of being forced to close operations due to social or environmental misconduct can be traced back to the eighteenth-century Anglo-Peruvian Amazon Rubber Co, also known as the Peruvian Amazon Company, which was listed on the London Stock Exchange (Velez-Ocampo et al., 2016). More recently, there is evidence of oil companies operating in Nigeria that have been found responsible for hiring private military security to open fire on peaceful protesters who, due to water depletion, have resisted oil extraction in the Niger Delta (Kalu & Ott, 2019; Taleb, 2017; Zalik, 2004). Likewise, the extraction of deep well water by Coca-Cola caused water shortages in many communities as well as destroyed local agriculture (Berglund, 2017; Ciafone, 2019). In India, Monsanto created a dependency cycle as its herbicide glyphosate accumulates toxins in the soil, which eventually became infertile. Farmers are thus forced to purchase genetically modified seeds that resist the herbicide (Glenna & Bruce, 2021; Rebière & Mavoori, 2020). These types of actions and omissions put corporations under public and governmental scrutiny, which risks their reputation (Gomez-Trujillo et al., 2020).

A second motive for companies to engage in sustainable practices is to account for consumers' backlash and sustainability awareness. While several non-governmental organizations have steered consumers toward conscientious consumption in recent years, there is also a present philosophical perspective that buying decisions should reflect consumers' values (Kingston, 2021). To illustrate this, conscientious consumption may be defined as 'setting and following personal policies of shunning (or selecting) certain goods, concerning their apparent connection (or lack of connection) to grave flaws upstream in the supply chain' (Kingston, 2021, p. 243). In addition, more consumers are becoming concerned about the environment and human rights, as there is an increasing awareness of the production dynamics and the impacts of responsible consumption.

A third motive is brand revitalization and public relations. There is evidence that consumer behavior of millennials (Y generation) and centennials (Z generation) is more affected by ethical beliefs than previous generations. As a response, there is market segmentation for sustainable productions. Young consumers are willing to compromise other product or service attributes (such as price) when consuming in reflection of their commitment toward sustainability. Therefore, there is an expectation for brands to tackle sustainability challenges. For example, according to a 2021 report published by Allied Market Research, the global green technology and sustainability market was valued at USD 10.32 billion in 2020 and is projected to reach USD 74.64 billion by 2030, with a compound annual growth rate of 21.9% from 2021 to 2030 (Allied Market Research, 2021).

Increasing corporate global value chain resilience can be another motive to adopt sustainability measures. The pandemic demonstrated that sustainability is not just limited to traditional social, environmental, and economic dimensions. Instead, sustainability and integrity are cross-cutting issues for large companies. COVID-19 has caused supply chain shortages that have even challenged national security in countries like the United States. Examples of such shortages would be semiconductors, advanced batteries, active pharmaceutical ingredients, and critical minerals.

A fifth motive may be to 'do well by doing good,' which was a key message by Benjamin Franklin to citizens, consumers, and companies to promote a balanced philanthropic and for-profit society. This message has increasingly been internalized and communicated in many companies worldwide (Hult et al., 2018). For example, Michael Porter and Mark R. Kramer (2011) describe this as creating shared value. In their own words, this can be articulated as 'policies and operating practices that enhance the competitiveness of a company while simultaneously advancing the economic and social conditions in the communities in which it operates' (Porter & Kramer, 2011). In addition to this, several business leaders have incorporated other elements into corporate decision-making, such as the idea of serving all stakeholders instead of focusing only on increasing shareholder value.

Scholars contend that socially and environmentally committed businesses emerge as responses to the environmental degradation and social inequalities associated with globalization (Bansal & DesJardine, 2014; Khan & Lund-Thomsen, 2011; Renouard & Lado, 2012; Scott, 2007; Swift & Zadek, 2002; Taylor, 2011).

For example, Hopkins (1999) proposes that reversing the adverse outgrowths of globalization is unavoidable for a 'planetary bargain' between the public and private sectors. This is supported by how in *Civil Society* by Michael Edwards (2004a), there is a suggested reciprocal relationship between economic actors and civil society. Edwards (2004b) in his book *Future Positive: International Co-operation in the 21st Century*, also asserted that no modern society can progress and sustain social aims without access to market economies.

Concerning business-related climate commitments, emissions are critical in quantifying scope 1 (direct GHS emissions from the company's direct operation) and scope 2 (emissions created during the production of energy used by the organization). Nonetheless, traditional profit-driven management had minimal engagement with supply chains and a merely transactional relationship. Therefore, traditionally, traceability was less important. However, by the third decade of the twenty-first century, traceability of the extended value chain's (downstream and upstream) impacts is a market entry requirement, an opportunity for eco-efficiency, and a means to multiply one's positive impacts. When measuring scope 3 (emissions resulting from assessments not directly controlled by the organization), companies can potentially send market signals to their suppliers via prioritization and market preferences to those more committed to reducing their GHG emissions.

The contribution of corporations to sustainable development in the territories where they operate has been documented, deeply questioned, mistrusted, but also frequently admired (Romani & Grappi, 2014). Likewise, since the end of the twentieth century, corporations have been more visible and, in many cases, more innovative and effective than the contributions of many local and national governments to sustainable development. This was largely motivated through corporate social responsibility or a 'commitment to improve community well-being through discretionary business practices and contributions of corporate resources' (Kotler & Lee, 2005, p. 3). Partnerships between businesses and governments are becoming increasingly important in terms of scope, relevance, and depth because the efficiency of service provision under all ownership models (public, private, or mixed) is highly dependent on financial and legal institutional development (GPSE, 2015).

Have Corporations Been Successful in a Way Governments Have Not?

There is no single answer to this question. Large companies, particularly multinational enterprises (MNEs), can influence their value chains upstream and downstream, and thus have the potential to exponentially accelerate SDG achievement. With an increasing demand to respond to regulators and investors, publicly traded firms must report and respond to ESG principles, actions, and standards. Furthermore, large companies committed to the SDGs can provide market incentives and sanctions to their suppliers. Incentives in terms of preferential access and prioritization are conditional to meeting ESG criteria, and there are sanctions in terms of excluding a supplier that has not met expectations. Businesses are also clients and consumers of services and goods. The private sector can send market signals that exponentially influence socioeconomic fabrics via business-to-business transactions. In a hypothetical example, Walmart's supply chain comprises 2,800 suppliers worldwide (Best & James, 2021). Each of those suppliers (Tier 1) has its own providers (Tier 2), who in turn have their own suppliers (Tier 3). It is increasingly expected that companies be held accountable for their multi-tier supply chain traceability. Hence, if Walmart commits to having a sustainable value chain that requires a positive ESG performance criterion for their Tier 1 suppliers, it will create a domino effect for suppliers to include these requirements throughout the value chain.

Secondly, unlike national and subnational governments, MNEs do not have an exclusive jurisdiction that limits their reach of demanding accountability. For instance, according to the Analytical Database on Individual Multinationals and Affiliates (ADIMA, 2021), Unilever PLC has its headquarters in the United Kingdom, with 696 affiliates and a direct physical presence in 109 jurisdictions. Likewise, the Danish AP Moeller-Maersk A/S has a direct physical presence in 109 jurisdictions, and Deutsche Post AG has 867 affiliates and a direct physical presence in 154 jurisdictions. It implies that if a large MNE adopts ESG criteria within their strategic decision-making and increases their attention to ESG data, their corporate strategy will have global outreach. As stated in the speech of Ms. Mia Mottley, Prime Minister of Barbados, at COP26 in November 2021, 'national solutions to global problems do not work.' While this may be due to market power, managerial capacity, and global outreach, it remains that MNEs can provide a unique and solid international leadership in the implementation of the SDGs (van Tulder et al., 2021).

Thirdly, there is an increasing presence of conscious consumers and ESG investing, which leads to raised expectations for public value generation and the prominence of corporate citizenship. There are several instances where people distrust the government, especially when there is a failure to fulfill the social contract or issues with the delivery of services. For example, in relation to taxes, there is a common sentiment associated with the suspicion of the government misusing taxpayers' tributes (Banerjee & Duflo, 2019). During the first year of the pandemic, there was an awareness of the monumental importance of corporate citizenship that was increasing in public value, which can be attributed to both the pandemic and manifestations of social discontent (Ahmed et al., 2021; Gonzalez-Perez et al., 2021; Patuelli et al., 2021; Rivo-López et al., 2021; Theis & Daniels, 2020). Survival during the pandemic required being a corporate citizen. It meant listening to community needs; understanding environments, and identifying, respecting, and valuing communities' uniqueness in order to sustain effective relationships with societal actors. As a result, several companies worldwide maintained solidarity with such needs and reincorporated philanthropy into their sustainability models. For example, the Bill & Melinda Gates Foundation focused the foundation's work on dealing with COVID-19 (Theis & Daniels, 2020). By May 2020, corporate philanthropic funding worldwide in response to COVID-19 reached USD 10.2 billion (Parks, 2020). As a result, it can be safely said that organizations supported suffering communities. The private sector needs to encourage corporate volunteering as well as commit to concrete and transcendental actions in coordination with national and local governments, other companies, unions, and CSOs.

Lastly, global, national, and local private sector alliances and partnerships transform the mindset surrounding business leadership. The Triple Bottom Line's (people, planet, and profit) business consideration is gaining prominence (Elkington, 2018). This has influenced corporate accounting, such as through the development of different models of social return of investment (SROI), as seen in the True Value, Total Impact Management, Return on Social Investment (ROSI), and the creation of the Sustainability Accounting Standards Board (SASB). This is further present through initiatives such as the Global Reporting Initiative (GRI) and the Dow Jones Sustainability Indexes (DJSI). Economics and politics are gradually becoming more driven by values instead of solely by traditional economic values. For example, in August 2019, almost 200 CEOs of the largest companies in the United States adopted the Business Roundtable Statement on the Purpose of a Corporation, declaring that for companies to succeed over the long term, they need to consider the interests of all their stakeholders; deliver value to customers, invest in employees, deal fairly and ethically with suppliers, support the communities in which they work, and generate long-term value for shareholders.

The Power of Business Alliances

In recent decades, different corporate global alliances and initiatives have been consolidating around sustainable development. Private sector initiatives can help meet the needs of local communities and develop local territories. In some cases, they can even make up for the lacking state presence. Since the end of the past century, there have been several instances of global alliances created to meet development goals. One such example is the World Economic Forum (WEF). Initially, the European Management Forum, the WEF, was also founded around the early idea of stakeholder capitalism. Management of enterprises serves all stakeholders and interest groups, acting for their benefit to achieve long-term sustained growth and company prosperity (WEF, 2019). This is the essence of the World Economic Forum's 1973 manifesto, which was updated in 2020 to include a much-needed commitment to diversity, and to mark suppliers as partners in value creation, thus giving supply chain management greater relevance (Schwab, 2019a, b)

As a result of the WEF platform for cooperation, over 274 projects are currently active regarding technology, innovation, climate change, diversity, mobility, and more. Furthermore, the central topic of the 2020 Davos annual meeting was "Stakeholders for a Cohesive and Sustainable World." To boost this momentum, the WEF gathered over 160 "Lighthouse Projects," which constituted an extensive but not comprehensive list of private, public, and WEF initiatives that demonstrate coalitions of stakeholders that drive positive impact quickly, cost-effectively, while being scalable and acting as proof of global corporate citizenship and stakeholder capitalism. These projects include work on equality, inclusion, economic development, ecology, climate and food systems, technology, corporate citizenship, and geopolitics, which all relate to the Forum's strategic imperatives for 2020 (WEF, 2020).

Examples of these lighthouse projects are the Global Alliance for Youth, an alliance of companies with the shared ambition to help 6 million young people build employability skills for the future. The initiative began with Nestle in 2013 and now partners with 22 other like-minded international companies alongside the support of the World Bank (WEF, 2020). Another example is the Hemayet Lead Project. With the support of Women's World Banking, Swiss Re and AXA Egypt provide insurance to over 200,000 women entrepreneurs and their small businesses in Egypt.

Another recent initiative at the Conference of the Parties (COP26) by the WEF in partnership with the U.S. Special Presidential Envoy for Climate and over 30 global businesses would be the First Movers Coalition. This project aims to invest in innovative green technologies to make them available for massive scale-up by 2030 and enable net-zero emissions by 2050 at the latest, which entails working with the Mission Possible partnership and other related initiatives (WEF, 2021).

Another well-known collective initiative is the United Nations Global Compact (UNGC), which was convened in the 1999 WEF framework in Davos by Kofi Annan, the then Secretary-General of the United Nations, in order to support the achievement of the MDGs. By the year 2000, approximately 40 large companies and several social and community leaders came together with the nine principles of the Global Compact, which account for human rights, labor rights, and environmental protection (Gonzalez-Perez & Leonard, 2015; Kell, 2015). In addition, years later, the anti-corruption principle was added to the Global Compact.

Two decades later, former U.N. Secretary-General Ban Ki-Moon commended the Global Compact for its leadership: 'The business community has an important leadership role here. Not only can it help shape government commitments through its strategies, but it can also help build a global consensus around mitigation and adaptation at a time when political leaders appear to be shrinking their responsibilities under the Paris Agreement Climate Change Agreement' (Global Compact, 2020). By October 2021, over 14,000 companies in 162 countries in nearly all economic sectors will have signed their commitment to the ten U.N. Global Compact principles (Global Compact, 2021).

Although more than 90% of UNGC participants have established principles and policies, this alone will not result in change. According to the 2020 UNGC's Progress Report, action is not matching up with policy in any of the four areas of the Ten Principles. For example, only 53% of UNGC participants claimed to train their employees in human rights, 23% conduct human rights risk assessments with suppliers or subcontractors, and 18% carried out internal human rights risk assessments. Concerning labor, 74% of companies had measures established to assess work–life balance, but the impact of such actions has yet to be studied. However, only 29% of the participants claimed to conduct labor rights impact assessments.

With regard to the environment, though 67% of participants committed to reducing GHGs, only 15% have targets approved by the Science Based Targets initiative. Also, even though 62% claim they do environmental impact assessments, only 18% of investment decisions take externalities into account.

Finally, in terms of anti-corruption measures, 56% of companies train their employees in anti-corruption practices, but only 25% conduct corruption impact assessments.

There are also country-based business alliance initiatives in the emerging world—for instance, the Colombian G12. The pandemic created a community of practice (CoP) where twelve of the largest companies[1] operating in Colombia, which primarily included multinational corporations, state-owned enterprises (SOEs), and institutions with private and mixed ownership. This generated over 282,000 direct and indirect jobs, involves the purchase of goods and services from over 127,000 suppliers, and currently accounts for 17% of Colombia's GDP (Gonzalez-Perez, 2021). This community of influential companies collectively agreed to increase their engagement with their stakeholders to minimize the negative impact of their operations and accelerate the achievement of SDGs along their downstream and upstream value chains. Each of these firms has aligned their business strategies with the SDGs and has publicly declared their intention to work together with the Colombian government to join public–private efforts toward achieving the goals. By

[1] Bancolombia (banking and financial services), Bavaria (drink and brewery—subsidiary of the Belgique AB InBev in Colombia), Ecopetrol (oil and gas), Enel (energy—subsidiary of the Italian ENEL in Colombia), Grupo Argos, Grupo Éxito (a retail—subsidiary of the French Casino groupe in Colombia), Grupo Nutresa (food), Grupo Sura (health and financial services), ISA (energy), Movistar (telecommunications—subsidiary of the Spanish Telefónica in Colombia), Postobón (brewery), and Terpel (oil and gas).

the end of 2021, all of these firms had the SDGs as part of their long-term strategy, sustainability key performance indicators (KPIs) on their top management scorecard, and SDG targets integrated into their core business.

Regarding the response to COVID-19, in addition to maintaining solidarity, these companies voluntarily presented their commitment to protecting the different natural endemic ecosystems, peacebuilding, decarbonization, and empowerment efforts, while including remote, rural, and vulnerable communities in the various Colombian territories. Importantly, this entails helping people who were hurt by conflict and making it easier for rural citizens to find employment and be part of society and the economy.

While the G12 in Columbia represents a success, development challenges not only continue to persist, but have also been aggravated by the pandemic. The work remains unfinished, and it necessitates not only the involvement of more companies, governments, academics, and civilians, but also the cessation of all counterproductive actions.

The Challenge of Demonstrating Impact: Measuring and Disclosing Businesses' Social, Environmental, and Economic Impacts

At the core of a businesses' responsibilities is measuring, transparent accountability, and impact reporting. Sustainability performance measurements and assessments support the management of value creation for stakeholders as an ethically motivated operation (Freeman, 2010). Yet, there has been a reported difficulty in adequately measuring the impact of companies' sustainability activities. This can be done in various ways, so that it can be recorded when an organization, be it a business or otherwise, makes a positive or negative contribution and aims to genuinely work toward the SDGs.

It is crucial to measure social and environmental impacts and identify the return on investments in this field (Serafaim, 2020). There is no denying that the interest and commitment of the private sector to sustainability has grown. However, it is important to note this growth is not necessarily altruistic in nature but rather because this helps companies increase their operational efficiencies, reduce risks, and save money. In addition, this growth is also motivated by how companies and investors have identified and quantified benefits to society, the environment, and the climate that affect financial results, as seen through externality analyses and the return on social value measurement methodologies. These methodologies include the social return on investment (SROI), Return on Sustainability Investment (ROSI) created by the Center for Sustainable Development at New York University, True Value created by KPMG, Total Impact Measurement & Management (TIMM) by PwC., Bien+ designed by EAFIT, and the Total Societal Impact (TIS) by the Boston Consulting Group.

ESG initiatives are not necessarily operated or monetized with the intention to translate into financial results. Though without monetization, ESG activities become additional projects, and companies and investors are less likely to realize the value of integrating sustainability into their operations. Also, when costs and benefits are given a value, the uncertainty surrounding them decreases, which helps guide businesses in making related decisions (Atz et al., 2021).

Annual sustainability reports are vehicles for communicating with stakeholders, strategic allies, and key players in value chains. Reports facilitate decision-making on material issues and monitor how sustainability is embedded in business strategies. For many, it is the prime directory of how to account for sustainability in corporate strategy. These reports also show the spending of an organization and its contributions to society and the environment over a year.

Nonetheless, reporting constitutes more than a business responsibility or being a good corporate citizen. It helps companies increase their value while promoting organizational learning, drives innovation, and provides incentives for a culture of transparency (Wynn & Jones, 2019). As such, it is crucial that an organization addresses and prioritizes its materiality. This refers to the issues that directly or indirectly affect the ability of an organization to create, preserve, or destroy its economic, social, and environmental value, as well as that of its stakeholders and the public (GRI Standards, 2021).

One of the most widely known reporting systems is the GRI Sustainability Standards. The GRI Standards is a system of interrelated universal, sectorial, and topical standards based on instruments such as the OECD's "Guidelines for Multinational Enterprises," and the U.N.'s "Guiding Principles on Business and Human Rights." These systems of standards are also used to report their assessments of how organizations' stakeholders could be or are affected by the operations and decisions of their organizations. This could also potentially inform investment decisions as it provides information about how organizations integrate sustainable development into their business models and strategies, and identifies current and future financial, reputational, and operational risks related to their sustainability impacts (GRI Standards, 2021).

Reporting standards require companies to report on their material topics. For each sector, the GRI identifies a list of material issues. For instance, there are 22 issues in the oil and gas sector, and among them are GHG emissions, climate adaptation, resilience, and transition, air emissions, biodiversity, waste, water, and effluents. Furthermore, each of these material ESG indicators is linked with the SDGs, as shown in Table 3.1.

SDGs as More Than Climate Action for Businesses

Often, the corporate sustainability narrative appears to prioritize environmental and climate challenges over poverty and inequality, social and human development, and governance. Achieving the targets for climate action is not enough. Still, taking action on climate change is imperative, as its impact risks the survival of businesses and the achievement of all of the other 16 SDGs, so climate action would enable the other goals to move forward. At the November 2021 COP26 in Glasgow, there was a consensus on the importance of businesses adopting a net-zero commitment. Although, achieving this by 2050, especially for high-emission sectors such as oil and cement, is highly ambitious. Plans to decarbonize include finding energy efficiency solutions, adopting renewable energy, capturing and sequestering carbon, storing energy, and using nature-based solutions for climate change adaptation.

Despite the undeniable urgency of climate change, achieving the SDGs is not solely reliant on climate action. While the implications of climate change are severe and threatening to the survival of many businesses and the viability of human and government activities, it does not address the grand challenges that face humanity as a whole. To reach the 17 SDGs, there needs to be an understanding of their interlinked nature.

The 2030 Agenda reflects these challenges, all of which have systemic elements across social, environmental, and economic dimensions. Despite this diversity, the 17 SDGs and 169 related targets are interconnected and their fulfillment is indivisible from one another. Also, it was conveyed by the International Science Council (2017) that science-based tools and analyses are required to make solid and coherent decisions toward

Table 3.1 Connections between ESG indicators in the oil and gas sector and the SDGs

	SDG 1 No poverty	*SDG 2 Zero hunger*	*SDG 3 Good health and well-being*	*SDG 4 Quality education*	*SDG 5 Gender equality*	*SDG 6 Clean water and sanitation*	*SDG 7 Affordable and clean energy*	*SDG 8 Decent work and economic growth*	*SDG 9 Industry, innovation, and infrastructure*
GHG emissions									
Climate adaptation, resilience, and transition	X						X	X	X
Air emissions			X						
Biodiversity						X			
Waste			X			X			
Water and effluents						X			
Closure and rehabilitation				X					
Asset integrity and critical incident management									
Occupational health and safety			X						
Employment practices	X			X	X				
Non-discrimination and equal opportunity				X	X				
Forced labor and modern slavery									

(continued)

Table 3.1 (continued)

	SDG 1 No poverty	*SDG 2 Zero hunger*	*SDG 3 Good health and well-being*	*SDG 4 Quality education*	*SDG 5 Gender equality*	*SDG 6 Clean water and sanitation*	*SDG 7 Affordable and clean energy*	*SDG 8 Decent work and economic growth*	*SDG 9 Industry, innovation, and infrastructure*
Freedom of association and collective bargaining									
Economic impacts	X				X				X
Local communities	X		X		X	X			
Land and resource rights	X	X							
Rights of indigenous peoples	X		X		X				
Conflict and security									
Anti-competitive behavior									
Anti-corruption									
Payments to governments	X								
Public policy									

	SDG 10 *Reduced inequalities*	SDG 11 *Sustainable cities and communities*	SDG 12 *Responsible consumption and production*	SDG 13 *Climate action*	SDG 14 *Life below water*	SDG 15 *Life on land*	SDG 16 *Peace, justice, and strong institutions*	SDG 17 *Partnerships for the goals*
GHG Emissions				X	X			
Climate adaptation, resilience, and transition			X	X				
Air emissions		X				X		
Biodiversity			X		X	X		
Waste			X		X	X		
Water and effluents			X		X	X		
Closure and rehabilitation		X			X	X		
Asset integrity and critical incident management		X			X			
Occupational health and safety								
Employment practices	X							

(continued)

Table 3.1 (continued)

	SDG 10 *Reduced inequalities*	*SDG 11* *Sustainable cities and communities*	*SDG 12* *Responsible consumption and production*	*SDG 13* *Climate action*	*SDG 14* *Life below water*	*SDG 15* *Life on land*	*SDG 16* *Peace, justice, and strong institutions*	*SDG 17* *Partnerships for the goals*
Non-discrimination and equal opportunity	X						X	
Forced labor and modern slavery							X	
Freedom of association and collective bargaining							X	
Economic impacts	X							
Local communities							X	
Land and resource rights		X					X	
Rights of indigenous peoples		X					X	
Conflict and security							X	
Anti-competitive behavior			X				X	
Anti-corruption							X	
Payments to governments							X	X
Public policy							X	

Source GRI Standards (2021) adapted by the authors

meeting the goals by 2030. This implies the need to understand synergistic relationships and potential trade-offs in order to define priorities and implementation strategies. Furthermore, it is essential to realize that the 17 SDGs may be 'wicked problems' (W.P.s) (van Tulder, 2018, p. 36). It is argued that the SDGs are 'systemic in nature, complexly interrelated and materialized at the interface between public–private and profit–nonprofit interests' (Van Tulder, 2018, p. 36). As such, they are wicked both by nature and design.

To elaborate, the wicked (or systemic, ambiguous, complex, symptomatic of another problem, and conflictual) matters for business regarding SDG 5: Gender equality, should be considered. This is the goal to "achieve gender equality and empower all women and girls." Ten of the 17 SDGs include gender-specific targets, with 22% of SDG targets being gender-specific (United Nations, 2019). Gender equality was originally part of the United Nations 2000–2015 MDGs, which is now SDG 5 in the 2030 agenda (United Nations, 2015a, 2015b).

According to Eden and Wagstaff (2021), in the context of management and international business (IB) literature, gender equality has typically been defined as the equal treatment of women and men in the workplace (Eden & Gupta, 2017; O'Brien et al., 2017; UNCTAD, 2014; UNCTAD, 2018; U.N. Economic Commission for Europe, 2019). Rational approaches based on a single discipline cannot handle the wicked problem of achieving workplace gender equality.

However, it is essential to recognize that even for firms, SDG 5 is more than workplace gender equality; it is also about the empowerment of women and girls. The goal of SDG 5 is equality between men and women in their rights and opportunities, their valuation and treatment, and their empowerment (i.e., the fostering of women's voice and agency). Whether measured as workplace gender equality or as women's empowerment, SDG 5 remains a wicked problem. As such, Eden and Wagstaff (2021) recommend a new evidence-based policymaking framework for MNEs: the SDG Materiality Matrix. This is based on four factors: quality of evidence, salience of the target to the company, actionability of the target by the firm, and ethicality of the target for the company. MNEs can then use this framework to select and implement policies for their preferred SDG targets.

The United Nations Global Compact (GRI), and the World Business Council for Sustainable Development (WBCSD), established the SDG Compass. This was built after incorporating feedback received through three consultation periods from companies, government agencies, academic institutions, and CSOs worldwide (SDG Compass, 2021). By November 2021, the SDG Compass identified 1553 business indicators currently used to measure and report business' contributions to the SDGs.

Besides business indicators, there exist numerous tools to assess the impact of companies on the SDGs. For instance, for SDG 1 (eradicating poverty), the Base of the Pyramid Impact Assessment Framework (BoP IAF) developed by The William Davidson Institute at the University of Michigan, is utilized to help assess the effects of business pursuits on low-income stakeholders. For SDG 6 (clean water and sanitation) and SDG 15 (life on land), the Global Water Tool (GWT) developed by the WBCSD supports the identification of corporate water risks and possibilities. Another example would be the GHG Protocol by WBSCD and the World Resource Institute (WRI) which helps diagnose, quantify, and regulate GHG emissions. There is also the Anti-Bribery Checklist by Transparency International, which acts as a self-monitoring instrument to determine a firm's anti-corruption policy, implementation, monitoring, and evaluation tools. Lastly, among many other strategies, there is how the Water Footprint Network developed Water footprint as an indicator of direct and indirect freshwater use by consumers and producers in relation to SDG 6 and 12 (responsible consumption and production).

The Circular Economy

A path for tackling the SDGs while maintaining profit can be found in the circular economy, making it an appealing option for corporate engagement. For instance, consider the circular economy in connection with SDG 12 (responsible consumption and production). According to the WBCSD (2017), the circular economy is a USD 4.5 trillion opportunity. Unfortunately, the conventional linear economic model dominates business, society, and research. However, there is a growing interest in the concept of the circular economy. As a result, there is also a greater need to address how such a model may be developed in a world of global value chains (Hofstetter et al., 2021).

The concept of the circular economy, as introduced in the 1990s by British economists David Pearce and R. Kerry Turner, has its origins in industrial ecology. This idea was articulated in order to explain the interrelationship of four economic functions of the environment: a source of resources, a sink, a comfort value, and a fundamental life support system. Industrial ecology promotes the minimization of resources and cleaner technologies as well as highlights the benefits of recycling waste to facilitate industrial symbiosis processes. Overall, the circular economy is based on the principles of the 3Rs: Reduce, Reuse, and Recycle.

Although, Professor Mikael Skou Andersen from the University of Aarhus in Denmark argued that the circular economy cannot sustain recycling. While the first recycling options are simple and have obvious benefits, as it increases the benefits become more difficult and costly to achieve. In other words, there are a limited number of circular options.

Andersen explains that, unlike the conventional economic system perceived as open and linear (produce, use, and waste), the circular economy proposes a closed system where resource use and waste are reincorporated into the system. The circular economy adopts the first law of thermodynamics, in which total energy and matter are constant on planet Earth and belong in a closed system.

As such, the circular economy not only makes environmental sense but offers opportunities for efficiency and productivity in design, purchasing, and waste management. Recycling, renting, leasing, reusing, sharing, and increasing product durability are also other potential sources of profit.

The Private Sector's Role Reducing Inequalities and Ending All Forms of Poverty

In June 2021, the impact of COVID-19 on global poverty was estimated by D.H. Mahler (and colleagues), which found that in 2020, the pandemic landed 97 million more people in poverty (Mahler et al., 2021). According to global poverty estimates, in 2021, there were 711 million people on Earth living on less than USD 1.90 per day (Mahler et al., 2021).

Although poverty alleviation is not viewed as a strategic priority for large businesses, there has been a corporate interest in poverty since the Industrial Revolution (van Tulder, 2008). The alleviation of poverty is not only a moral duty, but it makes for a sensible business case. For one, poverty has an immediate effect on restricting the market growth. When citizens experience a lacking disposable income, their expenditure is in turn limited, and their investment possibilities are nearly null, all of which restrains prospects for increasing human capital and hinders knowledge-based competitiveness, which are a few constraints among many others. Also, the rise of poverty and inequality increases insecurity, and incorporating security provisions in business operations decreases efficiency.

The SDG Compass (2021) identified 80 business indicators to measure the contribution of businesses to 'ending poverty in all its forms everywhere' (SDG 1). Table 3.2 shows ten business indicators to measure target 1.1. that claims 'By 2030, get rid of extreme poverty for everyone in the world. Right now, extreme poverty is defined as living on less than USD 1.25 a day.'

Table 3.2 Business indicators for SDG Target 1.1. By 2030, eradicate extreme poverty for all people everywhere, currently measured as people living on less than USD 1.25 a day

Business theme	*Indicator source*	*Indicator description*
Access to financial services	GRI G4 financial services sector disclosures	Access points in low-populated or economically disadvantaged areas by type
Physical and economic displacement	GRI G4 electric utilities sector disclosures	Approach to managing the impacts of displacement
Access to financial services	Women's world banking gender performance indicators	Average initial loan balance as a percentage of per capita Gross National Income (GNI), by gender
Electricity access	GRI G4 electric utilities sector disclosures	Average plant availability factor by energy source and by a regulatory regime
Electricity access	GRI G4 electric utilities sector disclosures	Average power outage duration
Access to financial services	Women's world banking gender performance indicators	Client retention rate, by gender
Disaster/emergency planning and response	GRI G4 electric utilities sector disclosures	Contingency planning measures, disaster/emergency management plan and training programs, and recovery/restoration plans

(continued)

Table 3.2 (continued)

Business theme	*Indicator source*	*Indicator description*
Economic inclusion	GRI G4 sustainability reporting guidelines	Describe policies and practices used to promote economic inclusion when selecting suppliers. Examples of financial inclusion may include suppliers owned by women or staffed by vulnerable, marginalized, or underrepresented social groups; and small and medium-sized suppliers
Earnings, wages, and benefits	U.N. Global Compact–Oxfam poverty footprint	Do large employers along the value chain and the more comprehensive company system negotiate wage and benefits standards through collective agreement?
Earnings, wages, and benefits	U.N. Global Compact–Oxfam poverty footprint	Does the company have a mechanism/policy/code that seeks to ensure that small-scale suppliers, smallholders, and/or distributors are paid a fair price for goods, services, and/or crops supplied to the company, enabling sustainable trade? If yes, provide details

Source Adapted by the authors based on SDG Compass (2021)

In recent years, especially after the work of C.K. Prahalad and Stuart Hart, there has been widespread recognition of the idea to create commercial opportunities for businesses while serving public purposes. In other words, it is the idea of fighting poverty with profitability (Prahalad, 2006). Prahalad's fundamental base of the pyramid (BoP) perspective implies that people living in poverty presents new market opportunities for business. For one, on a small scale, the agricultural and non-agricultural enterprises of the poor are often remarkably successful, though many of these businesses are too small to be efficient or achieve economies of scale (Banerjee & Duflo, 2007).

Is the World at a Point of No Return?

Mastercard issued a USD 600 million Sustainability Bond at a fixed rate of 1.90% in March 2021, alongside announcing in November of the same year that its net-zero date would be moved up from 2050 to 2040. According to them, proceeds from this bond 'will support significant progress already made towards maximizing carbon reduction, supporting environmental choices for customers, and fostering inclusive growth.' As such, investors then have the opportunity to follow the advancement of these efforts (Mastercard, 2021).

Also, in September 2021, a group of Chief Financial Officers (CFOs) of global companies such as BASF, Verizon, A.I. InBev, H & M, and PIMCO pledged $500 billion toward their commitment to the SDGs (Mashayekhi, 2021).

Companies in developed nations and emerging markets are increasingly proactive in adopting the SDGs and incorporating ESG issues into their business strategy, disclosure, and investment decisions. By 2020, 96% of the world's largest 250 companies (the G250) will report on their sustainability performance (KPMG, 2020). According to the "Responsible Business Trends Report 2021" by the consultancy firm SustainAbility Institute (2021), this proactiveness could be explained by a myriad of facts, and among them would be:

I. Growing mainstream interest in sustainability and improving social, environmental, and economic footprints;
II. Increased ESG investments since the outbreak of COVID-19;
III. Changing the regulatory landscape toward low carbon economies;
IV. Market momentum for net-zero commitments;
V. Growing awareness of business impacts and dependencies on nature, and;
VI. Accelerating digital transformation and addressing digital inclusion to boost business resilience;
VII. Recognizing businesses' role in the prevention, mitigation, and remediation of human rights in order to maintain their social license to operate and build trust within communities;

VIII. Establishing high voluntary sustainability practices and sharing knowledge and experience to shape policy, norms, and regulatory frameworks; and
IX. Increasing societal and market pressures on companies to disclose what they are doing to ensure they meet the needs of their diverse stakeholders.

Measuring the performance of companies with ESG criteria besides financial indicators may be challenging. As such, there are different mechanisms one can approach this with. For instance, when **the** DJSI launched in 1999, it was one of the first global indices to track selected sustainability-driven companies within the S&P Global BMI (S&P, 2021). Each year, over 10,000 firms worldwide were evaluated. SAM, an investment engineering firm and international asset manager, identified companies included in the annual DJSI based on their Corporate Sustainability Assessment (CSA). They also perform a business intelligence ESG risk and Media and Stakeholders' Analysis (MSA) that include controversies and issues that have reputational, or financial impacts on a company, such as human rights, corruption, fraud, labor disputes, and environmental disasters (S&P, 2021). Companies are categorized using the Global Industry Classification System (GICS) in the CSA process. Also, the DJSI ethically excluded the alcohol, tobacco, adult entertainment, nuclear armament, and gambling industries.

The Business Case for Investing in Sustainable Development

In 2011, Antony Bugg-Levine and Jed Emerson published the book *Impact Investing: Transforming How We Make Money While Making a Difference*. This text explained how for over a century, there was a belief that for-profit investments should focus only on producing financial returns, and in contrast, wealthy people who cared about social and environmental problems donated money in an attempt to solve these problems, with the alternative being to wait for government intervention (Bugg-Levine & Emerson, 2011a). Even though Höchstädter and Scheck (2015) claim there are differences in how to define impact investing, the central idea is that investors can look for financial returns while also solving social and environmental problems.

The term "impact investing" is not new. It was coined in response to perceived the heavy moral connotations of "socially responsible investment" or "ethical investment." However, the concept has since evolved from when it was first proposed by Bugg-Levine and Emerson in 2000. Seven years later, a group of investors convened, all of whom were interested in evaluating their capital's potential and real return through more than a passive financial, and wanted to exert a positive influence with their wealth. The phrase "impact investing," was then directly linked to the group's focused action and optimism. It also fit with the plans of investors who wanted to combine investing with helping communities.

There is evidence dating back centuries that the background of impact investing was an international movement where social and environmental commitment were expressed through market mechanisms. For example, in seventeenth-century England, a group known as the Quakers sought to align their investments and purchasing decisions with their values. In the United States nineteenth-century colonial era, congregations linked to the Quakers created businesses aligned with their social values and thus managed to finance religious communities. Bugg-Levine and Emerson (2011a) explained that these legacies influenced the environmental movement of the seventies, the anti-apartheid divestment campaigns in South Africa in the eighties, and recent movements of conscious consumption, responsible production, fair trade, and socially responsible investments.

A newer sentiment is the conviction 'that impact investors are profoundly optimistic about the role businesses can play in advancing the common good and the leverage social enterprises can achieve by employing financial tools' (Bugg-Levine & Emerson, 2011b, p. 11). In other words, impact investors see business practices and their related activities as 'a powerful force that can be harnessed for good, rather than a necessary evil that must be curtailed' (Bugg-Levine & Emerson, 2011b, p. 11).

The impact investing movement focuses on private equity and direct lending because of the unmatched power these investments have to generate social impact. However, evidence shows that while it's possible for impact investors to achieve social impact alongside market-rate returns, it is a difficult endeavor (Brest & Born, 2013). Recent research found that field impact investing is unique in six characteristics: (1) invested capital; (2) the degree of commitment to investees; (3) the selection process; (4) social and business outcomes; (5) reporting of results; and (6) the role of government (Agrawal & Hockerts, 2021).

Furthermore, there are hundreds of tools to measure the impact of investment. One of them is the Impact Reporting & Investment Standards (IRIS) produced by the Global Impact Investing Network (GIIN). It is used to estimate social, environmental, and financial progress, assess deals, and increase the reliability of the impact investing industry.

Bridging the SDGs and ESG Indicators

The SDGs provide institutional investors and companies with a framework for guiding and highlighting SDG-aligned investments or allocating resources to contribute to the United Nations 2030 Agenda. At the same time, the SDGs have brought to the forefront elements that were implicit in ESG standards and metrics by highlighting the environmental and social, as well as positive and negative externalities created by companies (Consolandi et al., 2020).

The SDGs were launched in 2015, however, investors have for decades incorporated screening mechanisms and indicators to identify environmentally and socially responsible investments, as well as made informed investment decisions by taking into account the direct impact on the environment and the communities where they operate. This is supported by how in 2006, the UN Principles for Responsible Investment included ESG issues.

Strengthening mandatory requirements for the disclosure by firms on ESG criteria is one tool for reaching the SDGs (Plastun et al., 2020). Since the global financial crisis, the emergence of regulatory disclosure systems based on ESG criteria and SDGs has considerably grown (Plastun et al., 2020). According to the Principles for Responsible Investment (UNPRI, 2016), over 300 measures, both mandatory and optional, have been implemented in several countries.

The Only Social Responsibility of Business Is to Be Sustainable

As evidenced in this chapter, contributing to sustainable development is not merely an act of philanthropy for businesses, nor is it simply a response to the pressure of legal regulations or stakeholders' expectations. Instead, it is a matter of survival, reputation, legitimacy, and competitiveness.

The relationship between profit, sustainability, and CSR, respectively, has long been studied. However, CSR and sustainability are not one and the same, despite how policymakers, business leaders, activists, and academics may use the terms interchangeably (Gonzalez-Perez, 2013). Each has distinctive features and historical contexts and is informed by different paradigms. For one, sustainability requires a long-term perspective and trade-offs across time, while CSR is the social, environmental, and economic initiatives that businesses voluntarily take responsibility for (McWilliams & Siegel, 2001). While business ethics and moral imperatives influence CSR, moral imperatives do not dictate what a company's actions 'should or should not do for sustainability' (Bansal & DesJardine, 2014, p. 72). While the perception of doing good measures the value of CSR, sustainability is measured by societal, environmental, and economic benefits relative to their cost. Furthermore, CSR is mostly driven by outside stakeholders (communities, NGOs), whereas sustainability is motivated by business strategy.

It has often been assumed that the pressures of stock markets drive corporations to focus on quarterly results, and that this focus on short-termism is at the expense of environmental and social justice. As Beslik (2017) claimed in an article published by the World Economic Forum: 'The finance world's short-termism will destroy our communities, economies and the planet.' However, the existing evidence does not demonstrate this.

Roe (2022) considers that it is widely assumed without evidence that the negative impacts of companies on climate, biodiversity, or labor relations are consequences of companies sacrificing future value for short-term gain, which he refers to as a problem caused by the timing of 'when' a corporation pays. However, Roe (2022) also argued that the bigger problem is not 'when,' but 'who.' This refers to those negatively affected by the intentional behavior of a corporation that deliberately acted at the cost of harming others to increase its profit. This means short-termism is not to blame for the negative impacts of business on society or nature, but rather, that it is sourced in profit pressure.

In 1970, the essay *The Social Responsibility of Business to Increase its Profits*, by the Nobel Prize-winner Milton Friedman, clearly states, 'only people have responsibilities. A corporation is an artificial person and, in this sense, may have artificial responsibilities, but "business" as a whole cannot be said to have responsibilities, even in this vague sense' (Friedman, 1970). For over five decades, this essay provoked polarized reactions between those who consider social business exclusively about making profits and those who believe companies have additional responsibilities to stakeholders besides shareholders.

Since the end of the twentieth century, several economic and management studies have found an empirical correlation between a firm's commitment to social and environmental issues and financial performance (Atz et al., 2021; Carter, 2005; Chen, 2011; Galbreath, 2009; Gomez-Valencia et al., 2021; Gray & Smeltzer, 1989; McWilliams & Siegel, 2000, 2001). Ultimately, if companies genuinely committed to ESG factors do well, we as a global society would also benefit. Large companies employ a massive workforce, and small businesses are included in their global supply chains provide livelihoods to billions. The private sector generates dynamism and prosperity in organizations and people that act as suppliers, customers, and users throughout their upstream and downstream value chains. Several businesses reach traditionally socioeconomically excluded territories and are significant contributors to public spending. However, it is also worth noting that many of these companies' shareholders are not individuals, but rather public and private pension funds that invest in them. This supports how generating shareholder value allows companies to finance pensions and contribute to the welfare of society.

Concluding Remarks

Businesses and communities are at risk from both global threats and local environmental challenges. The response of corporations to sustainable development challenges cannot reduce this risk but should generate opportunities to increase the positive externalities on society and natural ecosystems.

Large and small businesses are without a doubt increasingly sensitive to the sustainable development needs of each specific location where they directly operate, as well as what they can influence in their downstream and upstream value chains. While businesses are involved in ESG innovation concerning SDG implementation, there is a need to fast-track progress toward the goals. The world is constantly changing and global challenges could rapidly worsen. As such, companies must rely on innovation and technology as SDG accelerators.

A given organization may have positive and negative externalities, and possibly make individual deliberate efforts toward positive social and environmental contributions. However, effective change comes from partnerships with other companies, CSOs, and the public sector. It is doubtless that through collaboration with national governments and municipalities, the private sector's contribution to sustainable development and value creation has massive potential in scope and depth.

The first year of COVID-19 and the magnitude of the climate crisis raised expectations and emphasized the urgent need for businesses to engage in strong partnerships with national and local governments to do more (Gonzalez-Perez, 2022). The pandemic incited questions regarding how to define what constitutes as global, national, local, value-added, and even value itself.

Aligning markets with the SDGs and boosting companies' ESG factors makes financial sense. A strong ESG approach generates new market opportunities and efficiency benefits, spurs innovation, and boosts reputation (Business & Sustainable Development Commission, 2017). Companies with better ESG scores attract and retain employees, customers, B2B clients, investors, and acquire their social license to operate. As such, they also provide better shareholder returns.

Indisputably, the private sector is critical in accelerating the pace toward the 2030 Agenda. Businesses form the market's core, but they simultaneously have a leading role in state and civil society dynamics. Ultimately, the key factors of accomplishing the SDGs include alliances with suppliers, investors, and consumers, collaborative endeavors with competitors, and partnerships with governments and civil society.

Finally, it must be emphasized that the SDGs were designed with governments in mind, not the private sector. Businesses can align their voluntary commitments and strategy to the SDGs and significantly contribute to accelerating and scaling up the 2030 Agenda by incorporating ESG factors into their strategy. However, while the private sector can contribute to the 2030 Agenda, they are not accountable in the same sense as national governments.

Acknowledgements The authors would like to express their gratitude to Maria Camila Vargas de la Hoz for her contributions to this chapter. The authors appreciate the efforts of Haajar Abu Ismail in editing this chapter.

References

ADIMA. (2021). Analytical database on individual multinationals and affiliates. *OECD*. https://www.oecd.org/sdd/its/measuring-multinational-enterprises.htm

Agrawal, A., & Hockerts, A. (2021). Impact investing: Review and research agenda. *Journal of Small Business & Entrepreneurship, 33*(2), 153–181. https://doi.org/10.1080/08276331.2018.1551457

Ahmed, J. U., Islam, Q. T., Ahmed, A., Faroque, A. R., & Uddin, M. J. (2021). Corporate social responsibility in the wake of COVID-19: Multiple cases of social responsibility as an organizational value. *Society and Business Review, 16*(4), 496–516. https://doi.org/10.1108/SBR-09-2020-0113

Allied Market Research. (2021). Green technology and sustainability market expected to reach $74.64 billion by 2030. *Press Release Allied Market Research*. https://www.alliedmarketresearch.com/press-release/green-technology-and-sustainability-market.html

Atz, U., Liu, Z., Bruno, C. C., & Holt, T. V. (2021, August 31). Does sustainability generate better financial performance? Review, meta-analysis, and propositions. Available at *SSRN:* https://ssrn.com/abstract=3708495 or http://dx.doi.org/10.2139/ssrn.3708495

Banerjee, A., & Duflo, E. (2007). The economic lives of the poor. *Journal of Economic Perspectives, 21*(1), 141–167.

Banerjee, A. V., & Duflo, E. (2019). *Good economics for hard times: Better answers to our biggest problems*. Public Affairs.

Bansal, P., & DesJardine, M. R. (2014). Business sustainability: It is about time. *Strategic Organizations, 12*(1), 70–78. https://doi.org/10.1177/1476127013520265

Berglund, H. (2017). Civil society and political protest in India—The case of Coca-Cola in Kerala. *India Review, 16*(3), 324–343. https://doi.org/10.1080/14736489.2017.1348086

Beslik, S. (2017). *The finance world's short-termism will destroy our communities, economies and the planet*. World Economic Forum (WEF). https://www.weforum.org/agenda/2017/03/the-finance-world-s-short-termism-will-destroy-our-communities-economies-and-the-planet/

Best, R., & James, M. (2021). 5 Key suppliers of Walmart. *Investopedia*. https://www.investopedia.com/articles/insights/050116/walmart-stock-analyzing-5-key-suppliers-wmt.asp

Brest, P., & Born, K. (2013, Fall). When can impact investing create real impact? *Stanford Social Innovation Review*, 22–31. https://community-wealth.org/sites/clone.community-wealth.org/files/downloads/article-brest-born.pdf

Bugg-Levine, A., & Emerson, L. (2011). *Impact investing: Transforming how we make money while making a difference*. Jossey-Bass.

Bugg-Levine, A., & Emerson, L. (2011). Impact investing: Transforming how we make money while making a difference. *Innovations, 6*(3), 9–18.

Business and Sustainable Development Commission. (2017). *Better business better world*. https://sustainabledevelopment.un.org/content/documents/2399BetterBusinessBetterWorld.pdf

Carter, C. R. (2005). Purchasing corporate responsibility and firm performance: The key mediating roles of organizational learning and supplier performance. *International Journal of Physical Distribution & Logistics Management, 35*(3), 177–194.

Chen, C. H. (2011). The major components of corporate social responsibility. *Journal of Global Responsibility, 2*(1), 85–99.

Ciafone, A. (2019). Water for life, not for Coca-Cola: Commodification, consumption, and environmental challenges in neoliberal India. In *Counter-Cola: A multinational history of the global corporation* (pp. 227–270). University of California Press. https://doi.org/10.1525/9780520970946-008

Consolandi, C., Phadke, H., Hawley, J., & Eccles, R. G. (2020). Material ESG outcomes and SDG externalities: Evaluating the health care sector's contribution to the SDGs. *Organization & Environment, 33*(4), 511–533.

Eden, L., & Gupta, S. F. (2017). culture and context matter: Gender in international Business and management. *Cross-Cultural and Strategic Management, 24*, 194–210.

Eden, L., & Wagstaff, M. F. (2021). Evidence-based policymaking and the wicked problem of SDG 5 gender equality. *Journal of International Business Policy, 4*, 28–57. https://doi.org/10.1057/s42214-020-00054-w

Edwards, M. (2004a). *Civil society*. Polity Press.

Edwards, M. (2004b). *Future positive: International cooperation in the 21st century* (Rev). Earthscan.

Elkington, J. (2018). 25 Years ago I coined the phrase "triple bottom line." Here's why it's time to rethink it. *Harvard Business Review*. https://hbr.org/2018/06/25-years-ago-i-coined-the-phrase-triple-bottom-line-heres-why-im-giving-up-on-it?registration=success

Freeman, R. E. (2010). Managing for stakeholders: Trade-offs or value creation. *Journal of Business Ethics, 96*, 7–9.

Friedman, M. (1970, September 13). The Social responsibility of business is to increase its profits. *The New York Times Magazine*. http://websites.umich.edu/~thecore/doc/Friedman.pdf

Galbreath, J. (2009). Building corporate social responsibility into strategy. *European Business Review, 21*(2), 109–127.

Glenna, L., & Bruce, A. (2021). Suborning science for profit. Monsanto, glyphosate, and private science research misconduct. *Research Policy, 50*. https://doi.org/10.1016/j.respol.2021.104290

Global Compact. (2020). *Ban Ki-moon addresses the U.N. Global Compact*. https://twitter.com/globalcompact/status/1272806150816313345?s=20
Global Compact. (2021). *See who's involved*. https://www.unglobalcompact.org
Gomez-Trujillo, A. M., Velez-Ocampo, J., & Gonzalez-Perez, M. A. (2020). A literature review on the causality between sustainability and corporate reputation. *Management of Environmental Quality: an International Journal, 31*(2), 406–430. https://doi.org/10.1108/meq-09-2019-0207
Gomez-Valencia, M., Gonzalez-Perez, M. A., & Gomez-Trujillo, A. M. (2021). "Six Ws" of sustainable development risks. *Business Strategy and the Environment*. https://doi.org/10.1002/bse.2794
Gonzalez-Perez, M. A. (2013). Corporate social responsibility and international business: A conceptual overview. *Advances in Sustainability and Environment, 11*, 1–35.
Gonzalez-Perez, M. A. (2021). Los esfuerzos que hacen las empresas generan gran valor a la sociedad. *Forbes*. https://forbes.co/2021/06/12/red-forbes/los-esfuerzos-que-hacen-las-empresas-generan-gran-valor-a-la-sociedad/
Gonzalez-Perez, M. A. (2022). *Regenerative and sustainable futures for Latin America and the Caribbean: Collective action for a region with a better tomorrow*. Emerald Group Publishing Limited.
Gonzalez-Perez, M. A., Cordova, M., Hermans, M., Nava-Aguirre, K. M., Monje Cueto, F., Mingo, S., Tobon, S., Rodriguez, C., Salvaj, E., & Floriani, D. (2021). Crises conducting stakeholder salience: Shifts in the evolution of private universities' governance in Latin America. *Corporate Governance, 21*(6), 1194–1114. https://doi.org/10.1108/CG-09-2020-0397
Gonzalez-Perez, M. A., & Leonard, L. (2015) *Beyond the U.N. Global Compact: Institutions and regulations*. Emerald Publishing.
Gonzalez-Perez, M. A., Mohieldin, M., Hult, T. G., & Velez-Ocampo, J. (2021). COVID-19, sustainable development challenges of Latin America and the Caribbean, and the potential engines for an SDGs-based recovery. *Management Research: Journal of the Iberoamerican Academy of Management, 19*(1), 22–37. https://doi.org/10.1108/MRJIAM-12-2020-1119
GPSE. (2015). *Is the private sector more efficient? A cautionary tale*. United Nations Development Programme – Global Centre for Public Service Excellence.
GRI. (2021). *GRI Standards update 2021*. Available at: https://www.globalreporting.org/how-to-use-the-gri-standards/gri-standards-english-language/
Gray, E. R., & Smeltzer, L. R. (1989). *Management the competitive edge*. Collier Macmillan.
Hammer, M. (2001). *The super efficient company*. Harvard Business School Publishing.

Höchstädter, A. K., & Scheck, B. (2015). What's in a name: An analysis of impact investing understandings by academics and practitioners. *Journal of Business Ethics, 132*(2), 449–475.

Hofstetter, J. S., De Marchi, V., Sarkis, J., Govindan, K., Klassen, R., Ometto, A. R., Spraul, K. S., Bocken, N., Ashton, W. S., Sharma, S., Jaeger-Erben, M., Jensen, C., Dewick, P., Schröder, P., Sinkovics, N., Ibrahim, S. E., Fiske, L., Goerzen, A., & Vazquez-Brust, D. (2021). From sustainable global value chains to circular economy—Different silos, different perspectives, but many opportunities to build bridges. *Circular Economy and Sustainability,* 1, 21–47. https://doi.org/10.1007/s43615-021-00015-2

Hopkins, M. (1999). *The planetary bargain: Corporate social responsibility comes of age*. Macmillan Press.

Hult, G. T. M., Mena, J., Gonzalez-Perez, M. A., Lagerström, K., & Hult, D. T. (2018). A ten country-company study of sustainability and product-market performance: Influences of doing good, warm glow, and price fairness. *Journal of Macromarketing, 38*(3), 242–261. https://doi.org/10.1177percent2F0276146718787017

International Science Council. (2017). *A guide to SDG interaction: From science to implementation*. https://council.science/wp-content/uploads/2017/05/SDGs-Guide-to-Interactions.pdf

Kalu, K. I., & Ott, K. (2019). Ethical issues in environmental pollution: Multinational corporations (MNCs) and oil industries in tropical regions—The Nigerian Niger-delta case. In M. Chemhuru (Ed.), *African environmental ethics: The international library of environmental, agricultural and food ethics* (Vol. 29). Springer. https://doi.org/10.1007/978-3-030-18807-8_18

Kell, G. (2015). Foreword. Pp. xiii–xiv. In M. A. Gonzalez-Perez, & L. Leonard (Eds.), *The U.N. Global Compact: Fair competition and environmental and labor justice in international markets*. Emerald Publishing.

Khan, F. R., & Lund-Thomsen, P. (2011). CSR as imperialism: Towards a phenomenological approach to CSR in the developing world. *Journal of Change Management, 11*(1), 73–90.

Kingston, E. (2021). Shopping with a conscience? The epistemic case for relinquishment over conscientious consumption. *Business Ethics Quarterly, 31*(2), 242–274. https://doi.org/10.1017/beq.2020.13

Korten, D. C. (1998). When corporations rule the world. *European Business Review, 98*(1). https://doi.org/10.1108/ebr.1998.05498aab.007

Kotler, P., & Lee, N. (2005). *Corporate social responsibility: Doing the most good for your company and your cause* (p. 3). Wiley & Sons, Inc.

KPMG. (2020). *The time has come: The KPMG survey of sustainability reporting 2020*. KPMG. https://assets.kpmg/content/dam/kpmg/xx/pdf/2020/11/the-time-has-come.pdf

Lishchuck, R. (2021). *How large would tech companies be if they were countries?* Mackeeper. https://mackeeper.com/blog/tech-giants-as-countries/

Mahler, D. H., Yonzan, N., Lakner, C., Castaneda Aguilar, R. A., & Wu, H. (2021). *Updated estimates of the impact of COVID-19 on global poverty: Turning the corner on the pandemic in 2021?* World Bank Blogs. https://blogs.worldbank.org/opendata/updated-estimates-impact-covid-19-global-poverty-turning-corner-pandemic-2021

Manyika, J., Birshan, M., Smit, S., Woetzel, J., Russell, K., & Purcell, L. (2021). *A new look at how corporations impact the economy and households*. McKinsey Global Institute. https://www.mckinsey.com/business-functions/strategy-and-corporate-finance/our-insights/a-new-look-at-how-corporations-impact-the-economy-and-households

Mashayekhi, R. (2021). Global CFOs pledge $500 billion toward U.N.'s 2030 sustainability goals. *Fortune*. https://fortune.com/2021/09/21/united-nations-cfo-taskforce-2030-sustainable-development-goals/

Mastercard. (2021). *Mastercard issues $600 million sustainability bond*. Mastercard Newsroom. https://www.nasdaq.com/articles/mastercard-accelerates-its-net-zero-timeline-to-2040-quick-facts-2021-11-01

McWilliams, A., & Siegel, D. (2000). Research notes and communications: Corporate social responsibility and financial performance: Correlation or Misspecification? *Strategic Management Journal, 21*, 603–609.

McWilliams, A., & Siegel, D. (2001). Corporate social responsibility: A theory of the firm perspective. *Academy of Management Review, 26*(1), 117–127.

Micklethwait, J., & Wooldridge, A. (2003). *The company: A short history of a revolutionary idea*. The Modern Library.

Miller, J. (2018). *The political power of global corporations*. John Wiley & Sons.

O'Brien, K., Fitzsimmons, T. W., Crane, M., & Head, B. (2017), Workplace gender inequality as a wicked problem: Implications for research and practice. In *Academy of management annual meeting proceedings*. Briarcliff Manor.

Parks, D. (2020). Coronavirus-related giving tops $10 billion worldwide. *The Chronicle of Philanthropy*. https://www.philanthropy.com/article/coronavirus-related-giving-tops-10-billion-worldwide/

Patuelli, A., Caldarelli, G., Lattanzi, N., & Saracco, F. (2021). Firms' challenges and social responsibilities during Covid-19: A Twitter analysis. *PLoS ONE, 16*(7), e0254748. https://doi.org/10.1371/journal.pone.0254748

Plastun, O. L., Makarenko, I. O., Khomutenko, L. I., Osetrova, O., & Shcherbakov, P. (2020). SDGs and ESG disclosure regulation: Is there an impact? Evidence from top-50 world economies. https://essuir.sumdu.edu.ua/handle/123456789/82717

Porter, M. E., & Kramer, M. R. (2011) Creating shared value: How to reinvent capitalism—And unleash a wave of innovation and growth. *Harvard Business Review*.

Prahalad, C. K. (2006). *The fortune at the bottom of the Pyramid: Eradicating poverty through profits*. Wharton School Publishing.

Rebière, P., & Mavoori, H. (2020). The Bayer-Monsanto fusion: Countering brand tarnishing and rebuilding reputation. *Journal of Business Strategy, 41*(1), 27–37. https://doi.org/10.1108/JBS-10-2018-0185

Renouard, C., & Lado, H. (2012). CSR and inequality in the Niger Delta (Nigeria). *Corporate Governance, 12*(4), 472–484.

Rivo-López, E., Villanueva-Villar, M., Michinel-Álvarez, M., & Reyes-Santías, F. (2021). Corporate social responsibility and family business in the time of COVID-19: Changing strategy? *Sustainability, 13*(3), 241. https://doi.org/10.3390/su13042041

Roe, M. J. (2022). *Missing the target: Why stock market short-termism is not the problem*. Oxford University Press.

Romani, S., & Grappi, S. (2014). How companies' good deeds encourage consumers to adopt pro-social behavior. *European Journal of Marketing, 48*(5/6), 943–963. https://doi.org/10.1108/EJM-06-2012-0364

S&P. (2021, April). Dow Jones sustainability indices: Index methodology. https://www.spglobal.com/spdji/en/documents/methodologies/methodology-dj-sustainability-indices.pdf

Schwab, K. (2019a). *Davos Manifesto 2020: The universal purpose of a company in the fourth industrial revolution*. World Economic Forum. https://www.weforum.org/agenda/2019/12/davos-manifesto-2020-the-universal-purpose-of-a-company-in-the-fourth-industrial-revolution/

Schwab, K. (2019b). *Davos Manifesto 1973: A code of ethics for business leaders*. World Economic Forum. https://www.weforum.org/agenda/2019/12/davos-manifesto-1973-a-code-of-ethics-for-business-leaders/

Scott, S. (2007). Corporate social responsibility and the fetter of profitability. *Social Responsibility Journal, 3*(4), 31–39.

SDG Compass. (2021). *Inventory of business indicators*. https://sdgcompass.org/business-indicators/

Serafaim, G. (2020, September) Social-impact efforts that create real value. *Harvard Business Review*. https://hbr.org/2020/09/social-impact-efforts-that-create-real-value

Silva, S., Nuzum, A.-K., & Schaltegger, S. (2019). Stakeholder expectations on sustainability performance measurement and assessment: A systematic literature review. *Journal of Cleaner Production, 217*, 204–215. https://doi.org/10.1016/j.jclepro.2019.01.203

Singer, A. E. (2006). Business strategy and poverty alleviation. *Journal of Business Ethics, 22*, 225–231.

SustainAbility Institute. (2021). *What's next for sustainable business? 2021 trends report*. Institute. https://www.sustainability.com/globalassets/sustainability.com/thinking/pdfs/2021/ermsi-2021trendsreport.pdf

Swift, T., & Zadek, S. (2002). *Corporate responsibility and the competitiveness of nations and communities*. The Copenhagen Centre, Accountability, Institute of Social and Ethical Accountability.

Taleb, W. (2017). Environment-related human rights violations in the Niger Delta in Nigeria. *International Journal of Humanities and Cultural Studies., 4*(3), 355–375.

Taylor, M. (2011). Race you to the bottom and back again? The uneven development of labor codes of conduct. *New Political Economy, 16*(4), 445–462.

Theis, M., & Daniels, A. (2020). How 25 big grant makers are responding to the Covid crisis. *The Chronicle of Philanthropy*. https://www.philanthropy.com/article/how-25-big-grant-makers-are-responding-to-the-covid-crisis/

U.N. Economic Commission for Europe. (2019, April 8). Towards a conceptual framework for measuring gender-in-trade in official statistics. In Note by the United Nations Conference on Trade and Development. Conference of European Statisticians, Work Session on Gender Statistics. *Working Paper No. 25*.

UNCTAD. (2014). Investment by TNCs and gender: Preliminary assessment and way forward. In *Investment for development policy research series*. UNCTAD.

UNCTAD. (2018, October). Better data and statistics for gender-responsive trade policy. *UNCTAD Policy Brief, No. 70*. UNCTAD.

United Nations. (2015a). *The millennium development goals report 2015*. United Nations. https://www.un.org/millenniumgoals/2015_MDG_Report/pdf/MDGpercent202015percent20revpercent20(Julypercent201).pdf. Accessed 2 July 2020.

United Nations. (2015b). Transforming our world: The 2030 agenda for sustainable development. In *Resolution adopted by the General Assembly on 25 September 2015b*. A/RES/70/1. U.N. General Assembly, Seventieth Session. Agenda items 15 and 116. United Nations.

United Nations. (2019). *Progress on the sustainable development goals: The gender snapshot 2019*. Department of Economic and Social Affairs, United Nations. https://www.unwomen.org/en/digital-library/publications/2019/09/progress-on-the-sustainable-development-goals-the-gender-snapshot-2019

UNPRI. (2016). *Global guide to responsible investment regulation*. https://www.unpri.org/policy/global-guide-to-responsible-investment-regulation/207.article

van Tulder, R. (2008). *The role of business in poverty reduction towards a sustainable corporate story?* UNRISD Flagship Report: Combating Poverty and Inequality. https://www.unrisd.org/80256B3C005BCCF9/percent28httpAuxPagespercent29/340A393D6E02EE97C12575E0002A6C62/$file/WEBvanTulII.pdf van Tulder, R. (2018). *Business and the sustainable development goals: A framework for effective corporate involvement*. Rotterdam School of Management, Erasmus University.

van Tulder, R., Rodrigues, S. B., Mirza, H., & Sexmith, K. (2021). The U.N'.s sustainable development goals: Can multinational enterprises lead the decade of action? *Journal of International Business Policy, 4*, 1–21.

Velez-Ocampo, J., Herrera-Cano, C., & Gonzalez-Perez, M. A. (2016). The Peruvian Amazon Company's death: The Jungle devoured them. *Advanced Series in Management, 15*, 35–46.

WEF. (2019). *The World Economic Forum: A partner in shaping history 1971–2020*. World Economic Forum. https://www3.weforum.org/docs/WEF_A_Partner_in_Shaping_History.pdf

WEF. (2020). *Stakeholders for a cohesive and sustainable world: The role of lighthouse projects*. World Economic Forum. https://www3.weforum.org/docs/WEF_Lighthouse_Project_Report.pdf

WEF. (2021). *First movers coalition is tackling the climate crisis*. World Economic Forum. https://www.weforum.org/our-impact/first-movers-coalition-is-tackling-the-climate-crisis/

Wynn, M., & Jones, P. (2019). Context and entrepreneurship in knowledge transfer partnerships with small business enterprises. *The International Journal of Entrepreneurship and Innovation, 20*(1), 8–20.

Zalik, A. (2004). The Niger delta: 'Petro violence' and 'partnership development.' *Review of African Political Economy, 31*(101), 401–424. https://doi.org/10.1080/0305624042000005512

CHAPTER 4

The Role of Local and Regional Governments in the SDGs: The Localization Agenda

Abstract Local and regional governments (LRG)s have a critical role to play towards the achievement of the sustainable development goals (SDG)s in conjunction with national governments, the private sector, and community-based organizations (CBO)s. While there are some instances of LRGs acting to integrate the SDGs into their plans and budgets as well as reporting on progress through Voluntary Local Reviews (VLRs), the reality is that there is still much to be done. For LRGs to make a meaningful contribution towards the achievement of the SDGs through the localization process, they must strengthen their mandate, access to resources, planning and implementation capacities, as well as their inter-institutional coordination. This is a large task, especially when considering how at the end of 2020 there were only 39 VLRs. However, when considering the acceleration of SDG implementation in terms of climate action in the seven years after the Paris Agreement, LRGs have proven themselves capable of significantly scaling up their contribution. A major step forward is needed, as there are currently only eight years to go. In order to meet this deadline, LRGs will need technical and financial help to align their plans with the SDGs, and to scale up their investments and policy reforms. This will also require incentives to reward performance and the provision of special attention to building the capacity of LRGs that are most in need.

M. Mohieldin et al., *Business, Government and the SDGs*,
https://doi.org/10.1007/978-3-031-11196-9_4

Keywords SDG · Voluntary Local Reviews (VLRs) · Localization · Decentralization · Local government · Agenda 2030

Introduction

The goal to achieve the SDGs by 2030 is at serious risk (United Nations, 2021). This was already the case before COVID-19, due to insufficient progress by national governments and the impacts of climate change (United Nations, 2019). Since then, the pandemic has only aggravated the problem and further distanced the prospect of achieving the SDG targets. The incurred massive health and economic impacts not only caused an upheaval to the global economy, but it triggered the first reversal in global poverty reduction trends seen since at least a decade (World Bank, 2020). At the same time, LRGs are critical stakeholders to the achievement of the SDGs due to their critical role in service delivery. In addition, their proximity to citizens implies a better understanding of community needs and priorities. As such, if the global community is ever to attain the increasingly difficult target of achieving the SDGs by 2030, an all-hands-on-deck approach is necessary, including the private sector, LRGs, and civil society. The localization of the SDGs through the coordinated and complementary action between national and local governments will be especially important in accelerating implementation within the remaining short timeframe.

What Are LRGs' Efforts Toward the Achievement of the SDGs?

As the level of government closest to the target population, subnational governments including the second and third tiers of government (which together comprise of provinces, states, municipalities, cities, towns, and villages), all have a key role to play toward the achievement of the SDGs. While national governments are traditionally responsible for matters requiring territorial integrity (e.g., defense, foreign policy, macroeconomic, and fiscal issues) and the implementation of international agreements, LRGs play a key role in all aspects of service delivery. This is on account of the subsidiarity principle, wherein LRGs' service delivery is carried out by the level of government closest to the population they serve as LRGs have the best understanding of a given people's needs and priorities. In addition to service delivery, LRGs also play an important role in economic development, particularly in enabling access to employment. They also act as an important source of public investment as they are responsible for an estimated 37% of total public investments globally, with the figure rising to 60% in OECD countries (OECD, 2020; UCLG, 2019).

More recently, especially in the past decade, LRGs have taken new leadership roles in issues that were the traditional territory of national government. These included matters such as climate action, where leading cities joined forces to create the C40 cities and climate change network, and when LRGs worked to mobilize the Paris Agreement. Another area LRGs are dealing with would be the refugee crisis, which prompted them to importantly offer sanctuary and provide advocacy in both national and international arenas. This was especially visible with the influx of refugees displaced by the Syrian crisis to cities in the Middle East—as seen in Lebanon, Jordan, and Turkey,—and also to some extent in Europe.

The gap in achieving several of the global development targets has become more pronounced on the international stage. In comparison, LRGs have emerged as champions of SDG implementation. An example would be the leadership of New York City in 2018,—where several LRGs signed a declaration committed to carrying out VLRs and reports on the SDG progress within their jurisdiction. The declaration centered on three commitments: (1) to assess how their existing strategies, programs, and performance indicators align with the SDGs; (2) to create a stakeholder forum for knowledge and information exchange following the SDG framework; and (3) to submit a VLR during the High-Level Political Forum (HLPF). A total of 22 LRGs signed the New York City VLR declaration in September 2019, and while there is no official count, the amount of signatory LRGs has been growing steadily since then. By 2020, UCLG reports that some 39 VLRs have been published, with the city of New York having two VLR reports in 2018 and 2019 (UCLG & UN-Habitat, 2020).

The VLRs showed LRGs' commitment and efforts to align their strategies and plans to the SDG framework and to create relevant stakeholder forums. Yet, the VLRs portrayed much heterogeneity among LRGs. A study by IGES (2020) analyzed the content of VLRs prepared by 15 cities and found that only four covered the 17 SDGs, namely Bristol (United Kingdom), Hamamatsu (Japan), La Paz (Bolivia), and

Mannheim (Germany). The two VLRs submitted by New York City in 2018 and 2019 each analyzed 6 different SDGs, covering a total of 12. Otherwise, the remainder of VLRs analyzed only discussed an average of 6–8 SDGs (See Table 4.1). This is not to say all 17 SDGs need to be included, as some are naturally unachievable at the city level. However, the overall lacking holistic coverage of the SDGs may be in part due to the then limited guidance available to LRGs on how to prepare VLRs (a gap that was later filled by UCLG, OECD, and other institutions). Overall, though, it should be noted that LRGs vary greatly across countries in terms of devolved responsibilities, resources, and capacity.

Despite the proactive leadership from LRGs in the context of VLRs, the reality is that most national governments have rarely actively involved their subnational governments in the SDG implementation process. This applies to a variety of areas, be it monitoring, reporting, or the prioritization and collaboration of interventions. According to UCLG, out of the 143 countries that submitted VNRs to the UN's High-Level Policy Forum (HLPF) since 2016, adequate consultations with LRGs were carried out in only 42% of the cases. Another 22% of consultations were deemed weak, while 27% had no consultations at all (UCLG, 2019). This reflects a criticism of the development process of SDGs as top-down in nature, with insufficient consultation with LRGs and the private sector (Cernev & Fenner, 2019).

In addition, the aforementioned study showed that the state of consultations with LRGs in the context of VNRs did not have a consistent upward trend over time. In 2019, an estimated 38% of cases involved adequate consultations with LRGs, as opposed to the 46% from the previous year. This suggests that much remains to be done to confirm LRGs as an important force toward the achievement of the SDGs (UCLG, 2019).

Why Is the Involvement of LRGs Critical for the Achievement of the SDGs?

While not all SDGs are within the remit given to LRGs, an estimated 60–65% of the 169 SDGs targets require their involvement, whether it is through direct intervention or via coordination with national governments (OECD, 2020; UGLG & UN-Habitat, 2020). Of the 17 SDGs, SDG 11 and 16 specifically cater to local government institutions, with their respective goals to 'make cities and human settlements inclusive,

Table 4.1 Scope of the voluntary local reviews

Year		*SDG*																	
		1	*2*	*3*	*4*	*5*	*6*	*7*	*8*	*9*	*10*	*11*	*12*	*13*	*14*	*15*	*16*	*17*	*Count*
2018	HLPF						X	X				X	X			X		X	**6**
2018	Shimokawa						X	X		X		X	X			X			**6**
2018	Toyama						X	X		X		X	X			X		X	**7**
2018	Kitakyushu					X		X	X	X			X					X	**6**
2018	New York						X	X				X	X			X			**6**
2019	HLPF				X				X		X			X			X	X	**6**
2019	Bristol	X	X	X	X	X	X	X	X	X	X	X	X	X	X	X	X	X	**17**
2019	Buenos Aires				X				X		X			X			X	X	**6**
2019	Hamamatsu	X	X	X	X	X	X	X	X	X	X	X	X	X	X	X	X	X	**17**
2019	Helsinki				X				X		X			X			X	X	**6**
2019	La Paz	X	X	X	X	X	X	X	X	X	X	X	X	X	X	X	X	X	**17**
2019	Los Angeles				X				X		X	X		X			X	X	**8**
2019	New Taipei				X		X	X	X		X	X	X	X		X	X	X	**11**
2019	New York				X				X		X			X		X	X	X	**6**
2019	Santana	X		X	X							X					X	X	**7**
2019	Taipei			X				X	X			X	X	X				X	**7**
2020	Manheim	X	X	X	X	X	X	X	X	X	X	X	X	X	X	X	X	X	**17**
2020	Oaxaca																		**0**
VLR Total		**5**	**4**	**6**	**10**	**6**	**7**	**10**	**10**	**7**	**9**	**11**	**10**	**10**	**4**	**9**	**10**	**13**	

Source Adapted from Ortiz-Moya et al., (2020). State of the Voluntary Local Reviews P.8

safe, resilient and sustainable' as well as to 'build effective, accountable and inclusive institutions at all levels' (United Nations, 2015, pp. 24, 28). LRGs also have roles to play in the other goals as well, including initiatives related to infrastructure and service delivery, such as water, sanitation, health, education, housing, and transport, all of which depend upon decentralization.

As such, the localization of the SDGs and understanding the role that LRGs have in the implementation, spatial, and territorial dimensions of each goal has gained momentum as an important platform to enable the achievement of the SDGs. In stark contrast to decentralization, localization refers to an approach to the implementation of developmental objectives that combines the benefits of both centralization and decentralization (Elmassah & Mohieldin, 2020). As such, localization is not about an "either or" situation, but rather about combining the benefits of centralization in terms of resources and capacity with that of aligning local priorities with national development strategies as well as the benefits from a decentralized approach that enhances accountability to local communities (Elmassah & Mohieldin, 2020).

LRGs have a critical role to play toward the achievement of the SDGs. The link between various SDGs and LRGs—or the scope for the localization of the goals—is elaborated on in Table 4.2.

There are several reasons for the importance attributed to the localization of the SDGs. Firstly, LRGs have a critical role to play in achieving the SDGs through their responsibilities involving service delivery (e.g., water and sanitation, housing, transport, education, health) and socioeconomic development (e.g., decent work, infrastructure, ending poverty, tackling inequality). While the latter may not be surprising, LRGs had remarkably important contributions to the SDGs that have typically been thought to belong in national or international domains (e.g., climate change, oceans, ecosystems). These contributions were enabled by the importance that LRGs placed on environmental sustainability. This is normal in OECD countries, where LRGs are responsible for 64 and 55% of public investment and spending on the environment and climate-related issues, respectively (OECD, 2020). At the same time, an increasing number of subnational governments in developing countries are engaging in issues such as climate change. For instance, 54 of the 96 cities that form the C40 membership come from developing countries. This includes 2 cities (Addis Ababa and Freetown) from low-income countries (in accordance with the World Bank's classification of a per capita income of $1,045

Table 4.2 Role of the LRGs in achieving the SDGs

SDG1: End poverty	With their service delivery and economic development mandates, especially access to water, sanitation, and jobs, LRGs play an important role in poverty alleviation. Their proximity to citizens could help improve targeting of poor and vulnerable groups and their investment in disaster risk reduction would help reduce the vulnerability and strengthen the resilience of the poorest
SDG2: End hunger	LRGs can support food security and nutrition through investment and policies to improve farmers' access to markets and local supply chains through rural–urban linkages and enhance child nutrition in schools
SDG3: Healthy lives	LRGs play a role through providing water and sanitation, investing in slum upgrading and safe mobility to reduce road accidents, providing access to basic healthcare services, and tackling air and water pollution
SDG4: Education	LRGs are typically responsible for primary education and early childhood development, in addition to contributing to technical and vocational training
SDG5: Gender	With their service delivery mandates, LRGs can tackle barriers facing women (water and sanitation particularly affect women including menstrual health). With their urban planning and design mandate, LRGs can also ensure public spaces and transport are safe and accessible
SDG6: Water and Sanitation	LRGs are typically responsible for providing access to water and sanitation and their solid waste management mandate also contributes to reducing water pollution; they are thus critical for the achievement of this goal

(continued)

Table 4.2 (continued)

SDG7: Energy	While few LRGs are responsible for energy policies (Helsinki is one exception), their mandate over building regulations and investments in the municipal building stock are key to promote energy efficiency, as with their transport policies and investments which can also reduce emissions
SDG8: Growth and decent work	Through their focus on local economic development including related to sustainable tourism and their service delivery mandate, LRGs contribute to growth and creating jobs including for women and youth
SDG9: Infrastructure	LRGs contribute to the delivery of physical and digital infrastructure services as well as strengthening the resilience of infrastructure and services within their territorial boundaries
SDG10: Inequality	Through their service delivery and economic development mandate, LRGs can contribute to reducing inequality of access to services and economic opportunities, coupled with fiscal policies that prioritize budget allocation to marginalized groups
SDG11: Cities	The main goal concerning urban LRGs who have a mandate over nearly all targets—land use planning, housing, slum upgrading, service delivery, urban transport, cultural and natural heritage, disaster risk management, climate resilience, public spaces
SDG12: Consumption and production	Through their planning, regulation, and procurement capacities, LRGs can contribute to sustainable consumption and production including through green procurement, incentivizing local supply chains, conserving energy and water, reducing emissions and waste, and promoting sustainable tourism

(continued)

Table 4.2 (continued)

SDG13: Climate change	LRGs especially cities and regions have emerged as champions for climate change mitigation (through land use planning for more compact growth, policies, and investments to promote public transport, energy efficiency, etc.) and adaptation (through policies and investments to reduce disaster risk and both exposure and vulnerability from floods, cyclones, landslides, etc.)
SDG14: Oceans	LRGs especially coastal cities have an important role to reduce pollution of oceans through investments in solid waste management, plastics reduction, improved sanitation, and regulation of polluting industries and activities
SDG15: Ecosystems	LRGs have a role in preserving ecosystems, nature, and diversity including planning compact urban growth to avoid sprawl, policies such as payment for ecosystem services (PES) and community engagement to incentivize environmental protection, and regulation and disincentives to protect biodiversity such as polluter pay approaches, and investments in nature-based solutions
SDG16: Peaceful and inclusive societies and institutions	Through an emphasis on participation/community engagement and by taking action to reduce crime and violence and emphasize accountable institutions, LRGs have a role to play in this goal
SDG17: Implementation and Partnerships	Through an emphasis on strengthening subnational finances, relying on data-driven evidence-based policymaking and partnerships, LRGs have a role to play. The Voluntary Local Reviews (VLR) is one example of LRGs' leadership

Source Authors compilation based on UCLG (2020)

or less), and an additional 17 cities from lower-middle-income countries (per capita income of $1,046 to $4,905): Abidjan, Accra, Dakar, Dar-es-Salaam, Lagos, Nairobi, Hanoi, Ho Che Minh City, Jakarta, Quezon City, Chennai, Delhi NCT, Dhaka North, Dhaka South, Karachi, Kolkata, and Mumbai.

Secondly, LRGs are central protagonists in key trends that have a major bearing on the achievement of the SDGs, such as urbanization and climate change. Because urbanization typically goes hand in hand with improved poverty alleviation and increased access to basic services such as water, sanitation, health, and education (World Bank & International Monetary Fund, 2013), this is likely to have an acceleration effect on several SDGs if implemented well, including on SDG1, 3, 4, 6, and 11. In the same sense, a failure to act decisively on climate change to ensure alignment with the Paris Agreement's 1.5° Celsius target would put the achievement of multiple SDGs at risk. This includes SDG1, 2 (as the variability of temperature and precipitation affects agriculture and food production), 5 (due to the disproportionate impact of climate change on women), 13, 15, and 16 (with the destabilizing impacts of climate-induced migration). LRGs have a central role to play, both through how cities will effectively accommodate the expected 2.5 billion population increase by 2050 as the world continues to urbanize, and toward climate action including mitigation and adaptation.

Thirdly, to accelerate the achievement of the SDGs, the adoption of a spatial and territorial lens gains importance in helping to uncover the prominent geographic gaps of each target. This would allow national governments to target the needed financial resources and capacity in support of concerned LRGs. Currently, due to the great heterogeneity in most countries between urban and rural areas as well as leading and lagging regions, national figures have little relevance in informing governments on how to efficiently direct public resources toward areas most in need of investment.

Spatial disparities, whether in regards to accessing services, welfare indicators or economic development—are found everywhere. This applies to wealthier OECD countries, where there are great disparities in the achievements of various SDGs between countries as well as between regions and cities. In this sense, the third chapter of the OECD report 'A Territorial Approach to the Sustainable Development Goals: Synthesis report,' presents a detailed analysis of LRGs' gap to SDG achievement (OECD, 2020). For most goals, the majority of regions and cities within OECD countries largely remain at a considerable distance from the 2030 target. In fact, the average distance to targets in lagging regions across OECD countries ranges at around 40%, with some SDGs,—especially SDG14 on life under water,—demonstrating particularly deficient implementation (OECD, 2020). Overall, the share of regions within OECD states that have not achieved the end target of various SDGs ranges between 80 and 100%. ICities demonstrate a better track record of implementation in comparison. The share of cities that have yet to achieve the end target ranges between 30 to a little less than 100%. The best performance concerns SDG2 on hunger and food security, which overall stands at a 30% distance from the target including 100% achievement in six countries (all from the European Union) (OECD, 2020). Across the board, it can also be said that urban areas have a better track record in implementing SDGs than rural locations.

In this regard, the SDGs can become a powerful tool to reduce regional inequalities within a country, employing the "leave no one behind" principle as a call for action. For example, the Japanese national government identified the revitalization of rural regions in the country as one of the country's main challenges, owing to a declining population and a lack of economic opportunities. The government then identified the localization of the SDGs as the means to advance its regional development policies. In order to do so, it set up the SDG Future Cities program and established a policy target of enlisting 60% of all local governments to achieve the SDGs by the year 2024 (Masuda et al., 2021). As a result, by 2021, some 710 local governments (about 41% of Japan's total municipalities) had taken some action toward SDG implementation (Masuda et al., 2021). A prime example is the city of Kitakyushu, a former industrial hub that suffered from severe pollution and has promoted sustainable development since the mid-1960s. By the 1990s, the city had already gained global recognition for its environmental sustainability. As one of the first participants in the SDG Future Cities program (which channels financial resources to local governments), Kitakyushu has incorporated the SDGs into its sectoral plans since 2017, including in areas related to the environment, gender, and education. More importantly, beyond this planning, the city's Mayor oversees a cross-sector committee that coordinates issues relating to implementation, the local council and has started linking the SDGs to specific budget programs (Masuda et al., 2021).

What Are the Main Issues Preventing Local and Regional Governments from Achieving the SDGs?

Despite the efforts that subnational governments are doing to contribute to the SDGs (and it remains important for them to continue doing so), the reality is that LRGs have much to do before they can meaningfully implement the SDGs. The issues hindering the achievement of the goals are accentuated at lower levels of government, which in turn have a more limited capacity. Importantly, there are also obstacles that prevent LRGs from living up to their potential.

There are several, well-documented issues with the SDGs (Cernev & Fenner, 2019; Zinkernagel et al., 2018). First, the definition of the necessary targets to achieve various goals differs greatly. Some targets are clearly articulated and apply across countries and by extension LRGs, making them measurable (such as Target 1.1 'By 2030, eradicate extreme poverty for all people everywhere' or Target 3.1 'By 2030, reduce the global maternal mortality ratio to less than 70 per 100,000 live births'). On the other hand, many targets are either difficult to measure and enforce, or fundamentally lack a clear definition. An example of a target with an absolute end value that is harder to measure and in turn implement is Target 2.1 'end hunger and ensure access by all people … to safe and nutritious food all year round.'

Other targets are relative, such as Target 1.2 'by 2030, reduce at least by half the proportion of men, women and children of all ages living in poverty in all its dimensions according to national definitions,' or Target 3.6 'By 2020, halve the number of global deaths and injuries from road traffic accidents.' While these may be measurable and achievable, they do not appear to take into account the cost–benefit ratio of the marginal dollar invested to help maximize the impact of scarce resources.

Another example of a target that lacks a clear definition is Target 3.9: 'By 2030, substantially reduce the number of deaths and illnesses from hazardous chemicals and air, water and soil pollution and contamination.' This is also true of Target 4.4: 'by 2030, substantially increase the number of youth and adults who have relevant skills, including technical and vocational skills, for employment, decent jobs and entrepreneurship.' Overall, there is confusion over whether the end target would meet the "substantially" test. The definition of targets would be left to local interpretation, which in a context of limited resources and competing demands could likely negatively affect the prioritization of ambiguous targets by LRGs.

Another example of unclear definitions is Target 1.5 'By 2030, build the resilience of the poor and those in vulnerable situations and reduce their exposure and vulnerability to climate-related extreme events and other economic, social and environmental shocks and disasters.' It does not specify an end value, and while the proposed results indicators (e.g., number of deaths per 100,000 persons, and direct economic loss as a percentage of GDP) are measurable, they do not help beyond showing the direction of travel (OECD, 2020).

A couple of noteworthy points would be the lacking relationship between various SDGs and the issue of planetary boundaries. Several countries among the highest achievers in the SDG Index (a composite index developed by Jeffrey Sachs to measure overall progress against the achievement of the SDGs), were proven to have material and carbon footprints far exceeding sustainable levels (Hickel, 2020). The interdependency between the SDGs is another little studied issue (Cernev & Fenner, 2019). The most critical challenge, however, is likely the top-down nature of the development process for the SDGs, with a present near-exclusive focus on the national level. This omits an important array of institutions critical for the achievement of the SDGs, such as LRGs, corporations, and communities.

The ability of LRGs to have a meaningful contribution to the achievement of the SDGs requires several conditions that are often absent or insufficiently present. Most importantly, strong local action requires a robust decentralization process that provides a clear devolution of responsibilities from the government to LRGs, coupled with the ability to mobilize and retain the resources needed to perform such job. Local governments in many developing countries lack authority over several core planning and service delivery functions. In Egypt, subnational governments do not have control over land usage planning or water and sanitation. A national authority and regional utilities perform these functions, respectively, which are both affiliated with the Ministry of Housing, Utilities, and New Urban Communities. In Tunisia, local authorities do not have control over water and sanitation, which are delivered by two utilities—SONEDE and ONAS,—which are affiliated with the central government (World Bank, 2017). Even within the C40 network, not all cities have the power over basic service delivery. In fact, only 77% of C40 cities have power over the water supply, while less than half (49%) have control over the energy supply (C40, 2022).Even solid waste management, which is traditionally one of the most frequent responsibilities assigned to LRGs, has only 82% of C40 cities with authority on the issue (C40, 2022).

There are also frequent cases where the devolution of responsibilities has not been accompanied by the empowerment of LRGs to mobilize the necessary resources to fulfill their mandate. For instance, in the city of Freetown, Sierra Leone's capital, there was an introduced revamped property tax system in 2020 to mobilize revenues for service delivery, only for the national government to stop the system's rollout (Sierra Leone's property tax, 2020). The absence or ambiguity of enabling regulatory frameworks governing city creditworthiness has been another factor preventing leading African cities such as Dakar and Kampala from borrowing from capital markets. The former due to conflict with the national government, and the latter because of a prohibited low debt ceiling that makes debt issuance uneconomic (Gorelick, 2019).

The issue of LRG capacity is also relevant. There is significant variation in any given country between LRGs in terms of planning and implementation capacity. Even in instances where the availability of financial resources might not be a constraint, LRGs that lack the capacity, tools, and human resources for planning and service delivery, will inherently face implementation challenges and have a limited absorptive capacity of available funds. This was the case in Kenya, after the introduction of a new devolution policy that channeled important resources from the central government to counties, with disproportionately higher per capita resource allocation to historically marginalized counties in the North and Northeast. Yet, the lacking capacity of such counties meant they were unable to implement their mandate and absorb available resources (World Bank, 2019). As such, beyond financial resources, LRGs with limited capacity will need significant support in investment planning and implementation so that they can attract the needed human resources (e.g., teachers, medical doctors, civil servants, etc.) and set up basic service delivery systems.

In addition to capacity building, institutional coordination on spatial planning and service delivery issues between LRGs within a metropolitan area or region is critical. Yet, this is typically one of the most complex hurdles to overcome due to a range of political economy issues. For instance, the delivery of some services such as solid waste management is more cost-effective if implemented at a regional level given the high construction costs of new landfills. Yet, the complex political economy of siting a new landfill, with the Not-In-My-Back-Yard (NIMBY) syndrome and disagreements over cost-sharing arrangements, often hinders regional cooperation. This is the case in Australia, where the prevalence of a competitive culture among local entities is one of the main trends that explains the difficulty in establishing inter-municipal cooperative partnerships for waste management (Tobin & Zaman, 2022). Similarly, public transport and land use planning activities typically need to be carried out at the metropolitan level across multiple municipal jurisdictions, though political differences over cost sharing between municipalities and the influence of existing interests may also result in suboptimal collaboration (Nieuwenhuijsen, 2020). In both cases, financial incentives provided by higher level governments, be it from national governments to regions, or from regional/provincial governments to municipalities, can overcome obstacles for collaboration. For instance, the State of Rio de Janeiro committed to finance priority metropolitan infrastructure investments in order to incentivize municipalities to cooperate in the preparation of the Structure Plan for the Rio de Janeiro Metropolitan Region (Gasparotto et al., 2020).

Another key issue is the scarcity of disaggregated, reliable data at the local or regional level to enable an understanding of the gaps in progress measurement of SDG achievement. Indeed, only a subset of SDG indicators has disaggregated data available at the regional and city level. In OECD countries, only 125 and 56 of the SDG indicators have relevant data available for regions and cities, respectively, representing 74 and 33% of the total 169 indicators for the SDGs (OECD, 2020). Another exercise in Brazil that aimed to construct a dashboard for Metropolitan Areas related to the achievement of the SDGs shows serious limitations related to data (Martinelli & Lindner, 2021). The dashboard could only focus on 8 indicators related to both social and environmental issues (SDGs 1, 3–7, 10, 12) for which municipal-level data could be found, and in these, 34 indicators were used of which only 15 either followed or were closely linked to UNSTAT indicators. The remaining 19 indicators drawn from the Brazilian Institute of Geography and Statistics (IBGE) were identified as relevant proxies adapted to the local context. Overall, the availability of municipal data was a serious issue; for some indicators (SDGs 2, 13, 14, and 15) no data could be found at the municipal level, while in other cases, the data available was significantly outdated (Martinelli & Lindner, 2021).

Another important concern when measuring the achievement of the SDGs at the city level is the misalignment between how cities are defined at the national level (which reflects administrative jurisdictions), and the reality on the ground of urbanized areas that typically extend over multiple jurisdictions. The latter dichotomy gave rise to the Functional Urban Area (FUA) approach that defines cities on the basis of the existing urban form/extent and prevailing commuting patterns. This approach—endorsed by the European Commission, OECD, and the World Bank—enables information regarding the spatial and economic dimensions of any city, which yields relevant results for aggregation and comparisons (Dijkstra et al., 2019).

Another problem is the lack of alignment between the SDGs and local development plans developed and implemented by LRGs. This is because there is lacking priority to align existing plans with the SDGs in a context of a constrained capacity and pressure to deliver quick results. Some cities and regions have embraced the SDGs and introduced a holistic territorial lens. For instance, Utrecht has aligned its plans with the SDGs and created a dashboard that enables municipal staff to compare their city's achievements with the global targets (Jones & Comfort, 2020). However, examples of LRGs that have aligned their plans with the SDGs are more the exception than the rule, especially in the developing world. This also applies to higher-income countries, with many LRGs yet to embrace the SDGs (OECD, 2020). For instance, 50% of Norwegian municipalities surveyed (mostly small and medium-sized) reported that the lack of resources and capacity were the constraint preventing their focus on the SDGs, while the Association of Flemish cities and municipalities reported that 35% of municipalities were not interested in sustainable development (OECD, 2020).

National governments have adopted different strategies to get LRGs to take needed action for the localization of the SDGs, such as mandating it via regulation. For instance, in Indonesia, a Presidential Decree issued in 2017 required LRGs to prepare SDG action plans and integrate the goals alongside the national medium-term development plan into their own regional and local medium-term plans (UCLG, 2019). To support the implementation, the Planning Ministry prepared technical guidelines and provided capacity building and support in data collection (UCLG, 2019). Other countries such as Pakistan have set up the institutional systems and platforms to strengthen vertical intergovernmental coordination for the implementation of the SDGs, though these efforts did not produce the intended results as they were deemed top-down and lacked the necessary capacity to act effectively (UCLG, 2019).

Several countries use the SDG framework to improve intergovernmental collaboration, especially between the national and local level. In particular, Germany has set up a national system to support the localization of the SDGs through a program called Municipalities for Global Sustainability. This has enabled the city of Bonn, among others, to put in place a local sustainability strategy focused on six key priorities for municipal action toward the SDGs (OECD, 2020). Similarly, in Japan, the updated SDG action plan of 2018 includes a pillar on regional action focused on localization, through which 10 LRGs received financial support from the national government to implement their SDG strategies (OECD, 2020). In Iceland, the national government involved the local authorities' association from the onset in its national working group on the SDGs, noting in its VNR that the achievement of the SDGs is a joint project of the state and local governments (UCLG, 2019). In these examples, a partnership of equals approach and financial incentives have enabled better intergovernmental coordination and maximized the efforts of LRGs.

The Impact of COVID-19 on the Achievement of the SDGs and the Responses of LRGs

COVID-19 overturned the global economy and disrupted economic activities, supply chains, and livelihoods everywhere. In the process, the pandemic caused a major global setback in terms of the achievement of the SDGs. For the first time in a generation, a setback in the global poverty alleviation trend occurred. An estimated 97 million people were pushed to extreme poverty in 2020, causing an 11.6% year-on-year increase in extreme poverty in comparison to several years of a steady 2–3% decline per annum by 2019 (See Fig. 4.1) (Lackner et al., 2020). The world population in extreme poverty was expected to improve in 2021 alongside the increase of vaccination rates and the reopening of the economy in several countries, but the global figures mask major inequalities and worsening poverty in Sub-Saharan Africa, the Middle East, and North Africa, as well as in Latin America and the Caribbean (Lackner et al., 2020).

The pandemic spread quickly across urban centers, particularly in overcrowded places like slums and informal settlements that lacked adequate

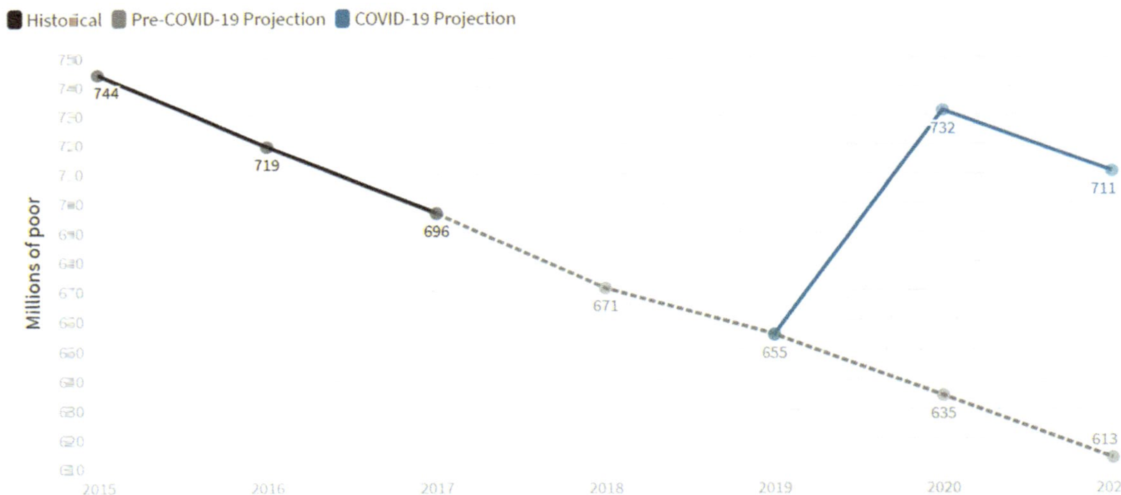

Fig. 4.1 Extreme poverty at the global level, 2015–2021 (in million population) (*Note* Extreme poverty is measured as the number of people living on less than $1.90 per day. 2017 is the last year with official global poverty estimates. Official poverty estimates are available for East Asia & Pacific, Europe & Central Asia, Latin America & Caribbean, and rest of the world for up to 2019, and for Middle East & North Africa and Sub-Saharan Africa up to 2018. Regions are categorized using PovcalNet definition. *Source* Lackner et al. [2020] [updated], PovcalNet, Global Economic Prospects)

housing, water, and sanitation infrastructure (Lall & Wahba, 2021). Additionally, cities faced a strong decline in economic activities as a result of government-imposed lockdowns to contain the spread of the disease, which was well documented through the intensity of nightlights that declined from pre-pandemic levels (Rubinyi et al., 2020). In the absence of compensation from the central government, the decline in economic activities translated into an important drop in cities' revenues during 2020 (mostly ranging between 10 and 30%) at a time when service delivery costs have increased to factor in public safety and service disruption considerations (National League of Cities, 2020; World Bank, 2022). In fact, the National League of Cities estimated a drop of $134 billion in city revenues due to the impact of COVID-19 (on tourism, real estate, and sales taxes, among others), with a 20.6% drop for the year (National League of Cities, 2020). As a result, cities' capacity to carry out capital investment in infrastructure and basic services had declined, in some cases by as much as 60% (World Bank, 2022).

At the same time, while those who could afford to work from home, many poor and vulnerable people among the 1.6–2 billion global workers in the informal sector without access to stable jobs, regular incomes, or safety nets, could not afford to stay home during lockdowns. This is supported by a study in Jakarta, Indonesia, which showed that the mobility of the poorest 10% of the city's population declined significantly less than the richest 10% (Fraiberger et al., 2020).

A combination of factors set back cities in SDG progress: drops in municipal revenues, capital investment in infrastructure, and basic service delivery, as well as declining economic activities, and the loss of jobs and livelihoods. This had a devastating impact on poor and vulnerable households. This decline especially applies to poverty and the delay of investments needed to attain SDG11 and the service delivery targets in multiple other SDGs.

However, COVID-19 appears to have at least impressed on several subnational authorities the importance of accelerating their efforts to put their LRGs on a pathway toward sustainability, while pursuing green, resilient, and inclusive development (GRID). There was a particularly detrimental loss of nature and biodiversity in cities, alongside the risks of increasing disasters and climate change. As such, several local governments have put a renewed emphasis on improving livability within their cities through facilitating active mobility (walking and cycling), expanding open spaces, and integrating nature. For instance, in 2020 Freetown launched a 3-year program to plant one million trees, with the aim of achieving multiple benefits: enhancing greening and tree coverage to tackle the urban heat island effect, strengthening resilience against flooding and soil erosion, and creating jobs. In fact, in its first year, the program enabled the creation of more than 500 employment opportunities for youth and community members responsible for planting and maintaining the trees (Tholley, 2021).

Bogota, Colombia's capital of 8 million inhabitants developed an ambitious plan in 2020 to increase its bicycle lane network by 50% throughout the city over a four-year timeframe, adding another 280 km to its 550 km network (Jaramillo, 2020). After the declaration of COVID-19 in March 2020,—the city added 84 km of emergency cycle lanes to its network as a way to offer alternative mobility options given the need to decongest the bus transit system that handled 40% of the city's daily trips. These emergency lanes were also later made permanent (Jaramillo, 2020). To incentivize the change, Bogota lowered the speed limit for cars, converted car parking to bicycle parking, implemented improved safety measures for cyclists, and reduced bike theft (Jaramillo, 2020). The result, when combined with the investment in public transportation, has been transformational in shifting people away from cars toward more efficient, green and inclusive modes of transportation, improving the mobility of everyone and especially the poor.

Many other cities including New York City, Washington DC, Paris, Barcelona, and Auckland reclaimed roads and converted them to widened sidewalks, bicycle lanes, and public spaces with outdoor dining areas for restaurants. In April 2020, the Barcelona city council approved a €4.4 million budget to pedestrianize several streets in the city center either partly or fully, while adding 21 km of 10 new bicycle lanes, as well as bus lanes and expanded sidewalks (Barcelona mobility plan, 2020). Milan proceeded to add 35 km of pop-up bicycle lanes and Paris even announced plans to add 650 km of bicycle lanes by 2024, as well as the aim to pedestrianize its iconic Champs Elysees Avenue (Reid, 2020). Both Paris and Milan later announced ambitious plans for a "15-minute city" wherein the day-to-day needs of the population (spaces for work, shopping, recreation, and education, among others) are to be available within a 15-minute commute from people's residences in order to minimize unnecessary commutes and maximize productivity, livability, and enjoyment of the city (Whittle, 2020). In addition, several cities—especially large Asian cities subject to recurrent flooding such as Bangkok,

Jakarta, Beijing, and Tianjin—are developing plans for investing in parks and green spaces, wetland restoration, and other nature-based solutions to integrate nature and reduce flooding risk.

Interventions to improve the quality of life in cities through an increased availability of public space through measures such as increasing mobility, pedestrianization, and greening have an immediate and visible impact without requiring major outlays of public funds. This is especially relevant given how the pandemic's impact translated into an important reduction on own-source revenues for cities coupled with an increase in the cost-of-service delivery, resulting in deep cuts of capital investment.

How Can LRGs Play a Meaningful Role Toward the Achievement of the SDGs?

LRGs have the ability to play a meaningful role toward the achievement of the SDGs if conditions are in place to maximize their potential impact. These consist of:

- Empowering LRGs via the decentralization of service delivery responsibilities to the level of government best suited to perform such functions. Large cities and metropolitan regions are typically ideal for carrying out spatial and land use planning, solid waste management, water supply and sanitation, public transportation, urban upgrading, as well as basic municipal social services such as primary education and basic health services. Although, tertiary education and specialized health facilities are typically the responsibilities of provincial or national governments. Smaller cities and municipalities may not have the economies of scale nor the capacity to carry out some of these services, which would benefit from a regional or metropolitan level solution. When LRGs have clear devolved responsibilities that are matched with adequate financial resources—including intergovernmental fiscal transfers—they are capable of delivering important progress toward the SDGs because they are best positioned to understand community preferences and priorities.
- Devolving responsibilities and the availability of adequate resources to implement LRGs' service delivery mandate. Robust local/regional finances depend on the ability of LRGs to generate

and retain own-source revenues, complemented with a transparent and equitable sharing of resources between local, provincial, and central governments. Additionally, to expand the finances available to LRGs, it is especially important to create the necessary conditions to attract private investment. This requires an enabling regulatory framework governing public–private partnerships (PPPs) in infrastructure and service delivery, as well as ground conditions such as a track record and local capacity for contract enforcement and implementation. The ability of cities to capture a portion of the increased value of land is an increasingly important financing tool for infrastructure and service delivery. This is due to government planning and infrastructure delivery decisions called Land Value Capture (LVC), such as the conversion of land from rural to urban, or a densification decision. An LVC can be applied ex ante through the sale of development rights or additional construction potential (as done in Sao Paulo and Rio de Janeiro, Brazil), or ex post through betterment levies (as seen in Bogota, Colombia).

- Strengthening LRGs' capacity to plan and deliver services while accommodating a growing population and attracting private investors and firms. However, not all LRGs have the capacity to carry out integrated spatial planning or to maintain service delivery at a level that matches population growth. Smaller LRGs and those located in lagging regions tend to have major gaps in terms of human resources that prevent them from carrying out their planning and service delivery mandates, even if they have the responsibilities and needed resources. The Upper Egypt Local Government Development Program financed by the World Bank invests in building the capacity of two governorates in Egypt's lagging region, which historically lacked resources and government capacity. This includes strengthening the capacity of local governments regarding participatory planning, citizen engagement to identify priority investments, the budget for these initiatives, and the ability to operate and maintain them (Grover et al., 2022).
- Aligning LRGs' administrative jurisdictions with the functional areas best suited for service delivery responsibilities. There often exists a disconnection between LRGs' administrative jurisdictions, which are previously set and infrequently adjusted, and the actual urbanized area that continues to grow over time. This often means

that rapidly growing urban areas expand beyond existing administrative jurisdictions and therefore service delivery responsibility for contiguous areas may fall under different municipal administrations. For example, the Sao Paulo metropolitan region comprises of 39 different municipalities—each one is responsible for land use planning, construction permitting, and sanitation services. As such, not all of Sao Paulo's municipalities use the state parastatal SABESP for sanitation services, which leads to diseconomies of scale (Ribeiro Lucinda, 2017). Having multiple municipalities responsible for service delivery within the same metropolitan region contributes to the additional challenge of having different planning and implementation capacities (Valencia et al., 2019). Moreover, the optimal scale for service delivery varies greatly from one issue to another. For instance, solid waste management involves multiple scales, as waste collection services occur at a local level, while disposal in a sanitary landfill is best carried out at a metropolitan scale (with one landfill typically servicing multiple municipalities). Public transportation is also best planned at a metropolitan level.

- Strengthening LRGs' resilience by increasing disaster and emergency preparedness. This is critical in ensuring that the progress achieved toward the SDGs is not jeopardized by future crises. LRGs that invest in disaster risk reduction (e.g., climate-proofing infrastructure, or risk-based land use planning and enforcement to protect areas at risk of flooding from unauthorized construction) bolster their capacity in emergency preparedness. This also includes strengthening their civil defence authorities, community awareness, and the preparedness of at-risk populations such as children and the elderly. Cities such as Buenos Aires and Cape Town, for one,—both members of the 100 Resilient Cities Initiative,—used their resilience strategies as the anchor development strategies with which to align their efforts for the localization of the SDGs (Valencia et al., 2019). Ultimately, investing in disaster risk reduction is critical for safeguarding development progress toward the attainment of the SDGs. In Bangladesh, four decades of continued investment in disaster risk management resulted in a drastic reduction in the loss of lives and property from the recurrent cyclones, which led to maintaining state efforts to alleviate poverty, strengthening human development, and contributions to economic and environmental sustainability. In 1970, Cyclone Bhola caused over 500,000 fatalities, making it the deadliest tropical cyclone ever recorded, in contrast, Cyclone Sidr of 2007 caused less than one percent of these fatalities despite having the same speed of about 223 km/hour (World Bank, 2019).

- Strengthening participation and the involvement of city residents in decision-making. This ensures that community priorities occupy center stage while mobilizing the electorate behind the SDGs. The city of Mannheim carried out extensive awareness building regarding the SDGs with its citizens while registering community priorities, and over a two-year period, the city came up with a Mannheim 2030 vision that is aligned with the implementation of the SDGs (Hubel, 2021).
- Access to technological platforms and solutions. This is important for both improved municipal management and service delivery, as well as for effectively monitoring the progress of implementation toward the SDGs. In the process of producing iVLR reports, Helsinki connected the SDGs to the city's strategy and created a whole-of-government approach to collect the data and track progress toward the attainment of the goals (Malin, 2021).
- City-to-city knowledge exchange and experience sharing about efforts to achieve the SDGs. United Cities and Local Government (UCLG), ICLEI Local Government for Sustainability, the C40, and Metropolis are among the most active city networks in promoting the SDGs. Also, organizations such as the World Economic Forum, OECD, UN-Habitat, and the World Bank are providing technical and financial support as well as analytics related to the achievement of the SDGs and the capacity building of LRGs.

Overall, and most importantly, the potential for scaling up action by LRGs would be complemented by central government actions and resources that are developed and budgeted alongside local communities' priorities and needs. The benefit of such a localization approach is that it pragmatically combines the benefits of both central and local governments in a coordinated approach that is aligned with local priorities. In the process, this creates a reinforced accountability toward local communities through budget alignments and interventions with their needs. It is these accountability measures that cause the channeling of additional resources from local communities and civil society toward the achievement of the SDGs, including some of the harder to reach populations and service delivery frontiers.

What Are the Roles of Local Communities and CSOs in Localizing the SDGs?

There is room for LRGs and CSOs to contribute beyond complementing the efforts of national governments to implement the SDGs. Some roles are particularly relevant for non-governmental entities, such as participating in defining priorities, and monitoring overall implementation and progress toward the SDGs. They also target specific constituencies such as marginalized groups, indigenous populations, minorities, and other vulnerable populations in order to "leave no one behind." In certain settings, especially in fragile, conflict-ridden states with an absence of governance, civil society, and community organizations take the lead with a proactive service delivery role regarding health, education, and basic services (Mohieldin & Shehata, 2021).

The role of communities is especially important. One particularly relevant approach is community-driven development (CDD), wherein the national government or international financial institutions pass on grants to local communities for the identification and implementation of community priorities including small-scale infrastructure projects. The CDD approach was praised for providing an alternative system for service delivery when local government capacity was lacking or public service delivery was perceived as corrupt and ineffective. It was also considered beneficial for its ability to channel funds to the poorest areas in a given country. On the other hand, community structures and social norms often dictate the outcomes. If participation is not accessible to all community members, or if the voices of certain community leaders or social groups dominate others, the resulting investments would not reflect the priority of the whole community. An evaluation of 23 CDD programs found that positive outcomes in terms of community infrastructure were linked to increased social cohesion or improved governance and decentralization (White et al., 2018).

Another relevant approach is the payment for environmental (or ecosystem) services (PES), wherein community members receive payments in return for protecting the environment or natural resources. In Quindio, Colombia, a PES scheme was adopted to encourage environmentally friendly land use changes through promoting biodiversity and the adoption of silvopastoral practices in degraded landscapes. An evaluation of the scheme showed that community members that received

PES outperformed the control group by 200% in terms of environmentally friendly land use conversion over a four-year timeframe, while contributing seven times as much environmental services (Pagiola & Rios, 2013). Similar arrangements were structured to incentivize farmers in the State of Espirito Santo in Brazil, in order to protect water resources from sedimentation and soil erosion that had obvious benefits regarding the quality and treatment cost of water resources. In Costa Rica, there was also the protection of a forest ecosystem conservation and prevention of land degradation, with over 18,000 families benefiting from the scheme to protect 1.3 million hectares of forest and land (United Nations, 2022).

Recently, there is a renewed optimistic sentiment about the potential for local communities and local organizations to improve the efficiency of development aid and achieve poverty alleviation results, be it through NGOs, CBOs, or social enterprises. The important role of NGOs and social enterprises in community-driven development has been established for several decades, especially since the emergence of micro-credit and the impact of Grameen Bank and grassroots organizations on poverty alleviation and community empowerment (Dongier et al., 2003; Wahid, 1994). Social enterprises gained even more importance during COVID-19 by extending critical relief, safety nets, and basic services to poor and vulnerable households in order to complement the responses of overwhelmed national and local governments (Das et al., 2020). These services were greatly varied and included efforts to spread awareness of COVID-19 and health precautions in Pakistan, the supplementation of water, sanitation, and hygiene (WASH) efforts in informal settlements in Kenyan cities, the development of online education tools and platforms in Jordan, and the provision of COVID-19 testing and affordable healthcare in Nigeria (Das et al., 2020).

These efforts did not go unnoticed by donors and philanthropies. USAID's administrator Samantha Power announced in November 2021 that at least one-quarter of their funds would be channeled to local partners over the coming four years, which represents a more than 400% increase from the current level of six percent (USAID, 2021). USAID recognizes that working with local organizations is difficult, time-consuming, risky and involves capacity and fiduciary challenges. However, the benefits in terms of reaching hard-to-reach areas, the strengthening of local capacity, and long-term sustainability impacts outweigh the costs (USAID, 2021). Similarly, new major philanthropic efforts are in place since the pandemic, including foundations such as the Bezos Earth Fund, or through individual donations such as MacKenzie Scott's distribution of billions of dollars to local organizations and NGOs. Development experts predict a continuation of the localization of philanthropic and development aid dollars channeled toward local organizations and social enterprises. This will be supported with enhanced accountability through technological advances, and the support of foundations in helping equip local organizations to better implement funds from development finance institutions (Kumar, 2022).

Concluding Remarks

There is much that LRGs can do to contribute to the achievement of the SDGs and complement the role of national governments, the private sector, and CBOs. There are numerous instances of interesting and proactive initiatives that LRGs have implemented, as seen with the VLRs. Notwithstanding this, for the localization of the SDGs to enable meaningful LRG contributions, there are several conditions that need to be met. These consist of strengthening the alignment of national and local government strategies, budgets, and actions with local communities' priorities and needs. In addition, this will require improving the accountability of central and local government institutions toward such needs, as well as reinforcing the devolution process in specific service delivery dimensions and accompanying it with adequate resources. All of which should be implemented while improving LRGs' planning, service delivery, disaster risk management, and emergency response capacity. Also, administrative jurisdictions should be aligned for service delivery in consideration of the spatial reality on the ground, and horizontal and vertical institutional coordination should be enabled across sectors. Although in practice, these conditions will likely be difficult to effectively carry out.

LRGs have already been demonstrating important leadership in stepping up to the climate crisis. To date, some 6,150 cities—members of the Global Covenant of Mayors—have already developed climate action plans (Negreiros et al., 2021). However, these cities are home to only 20% of the world's urban population, which means a great deal more needs to be done. In comparison, the situation concerning the SDGs is a different matter. Despite the importance of the role of LRGs, there is a demonstrably lower level of engagement or lacking awareness surrounding LRGs' efforts toward the SDGs. Currently present initiatives include LRGs' aligning their development plans with the SDGs, or reporting on their achievements toward the goals using the VLRs or other means. By end of 2020, only 39 VLRs were reportedly published, with some cities including New York having submitted more than one. Moreover, VLRs submitted to date have provided partial reporting on the SDGs, averaging six SDGs per VLR.

However, to expect LRGs to single-handedly compensate for the large and increasing gap toward the achievement of the SDGs would be a setup for failure. Rather, an all-hands-on-deck approach is needed in order to combine the benefits of central government action with the complementary resources and resourcefulness of LRGs, all the while ensuring the alignment of central and local strategies, initiatives, and budgets with accountability toward local communities.

With eight years left to the deadline of the 2030 Agenda, current progress is clearly insufficient. Therefore, a major step forward is critically needed. Such a step would require technical and financial support for LRGs to align their plans with the SDGs and scale-up implementation, the provision of incentives to reward performance, and the concentration of special attention on the capacity building of LRGs most in need.

Acknowledgements Authors recognize the support of Mg. Manuela Gomez-Valencia in formatting the chapter. They acknowledge as well Haajar Abu Ismail's excellent editorial contribution to the chapter.

References

C40. (2022). *C40 cities*. The Power to Act. https://www.c40.org/researches/c40-cities-the-power-to-act

CatalanNews. (2020, April 25). *Pedestrianised streets and wider pathways: Barcelona's plan for a return to mobility*. CatalanNews. https://www.catalannews.com/society-science/item/pedestrianised-streets-and-wider-pathways-barcelona-s-plan-for-a-return-to-mobility

Cernev, T., & Fenner, R. (2019). The importance of achieving foundational sustainable development goals in reducing global risk. *Futures, 115*. https://doi.org/10.1016/j.futures.2019.102492

Das, M. B., Khan, I. A., & Tinsley, E. (2020). *Results-based financing through social enterprises: A white paper for the global partnership for results-based approaches, in response to the Covid-19 pandemic*. Washington, DC:

World Bank. https://openknowledge.worldbank.org/handle/10986/34319. License: Creative Commons Attribution CC BY 3.0 IGO.

Dijkstra, L., Poelman H., & Veneri, P. (2019). *The EU-OECD definition of a functional urban area* (OECD Regional Development Working Papers, No. 2019/11). OECD Publishing. https://doi.org/10.1787/d58cb34d-en

Dongier, P., Van Domelen, J., Ostrom, E., Ryan, A., Wakeman, W., Bebbington, A., Alkire, S., Esmail, T., & Polski, M. (2003). *Community driven development*. World Bank Poverty Reduction Strategy. World Bank. https://web.worldbank.org/archive/website00518/WEB/PDF/CDD_CHAP.PDF

Elmassah, S., & Mohieldin, M. (2020). Digital transformation and localizing the sustainable development goals (SDGs). *Ecological Economics, 169*, 106490. Elsevier. https://doi.org/10.1016/j.ecolecon.2019.106490

Fraiberger, S., Jones, N., & Lozano Gracia, G. (2020). *Poor people respond differently to stay at home orders: Here's what data says*. The World Bank. https://blogs.worldbank.org/sustainablecities/poor-people-respond-differently-stay-home-orders-heres-what-data-says

Gasparotto, T., Ambrosano, J., & Brakarz, B. (2020). *Opportunities for sustainable infrastructure investments al city level in Brazil*. Clima Change Division. https://publications.iadb.org/publications/english/document/Opportunities-for-Sustainable-Infrastructure-Investments-at-City-Level-in-Brazil.pdf

Gorelick, J. (2019). Supporting the future of municipal bonds in Sub-Saharan Africa: The centrality of enabling environments and regulatory frameworks. *Environment and Urbanization*, February 2018. https://doi.org/10.1177/0956247817741853

Grover, A., Lall, S., & Maloney, W. (2022). *Place, productivity, and prosperity: Revisiting spatially targeted policies for regional development*. World Bank. https://doi.org/10.1596/978-1-4648-1670-3

Hickel, J. (2020). *The world's sustainable development goals aren't sustainable*. Foreign policy. https://foreignpolicy.com/2020/09/30/the-worlds-sustainable-development-goals-arent-sustainable/

Hubel, C. (2021). *Building Mannheim's SDG strategy around the vision of residents*. City playbooks for advancing the SDGs. Brookings Center for Sustainable Development. https://www.brookings.edu/wp-content/uploads/2021/07/City-playbook_Mannheim.pdf

IPCC. (2021). *Climate change 2021: The physical science basis*. Contribution of working Group I to the sixth assessment report of the intergovernmental panel on climate change. Cambridge University Press.

Jaramillo, A. (2020, August 10). *Bogota builds its future around bikes*. Bloomberg City Labs. https://www.bloomberg.com/news/articles/2020-08-10/to-tame-traffic-bogot-bets-big-on-bike-lanes

Jones, P. & Comfort, D. (2019). A commentary on the localization of the sustainable development goals. *Journal of Public Affairs, 2020*(20), e1943. https://doi.org/10.1002/pa.1943

Kumar, R. (2022). *Trends shaping global development in 2022*. Inside Development. Devex. https://www.devex.com/news/5-trends-shaping-global-development-in-2022-102375

Lackner, C., Gerszon Mahler, D., Negre, M., & Beer Prydz, E. (2020). *How much does reducing inequality matter for global poverty?* (Global Poverty Monitoring Technical Note 13). Word Bank Group. https://documents1.worldbank.org/curated/en/765601591733806023/pdf/How-Much-Does-Reducing-Inequality-Matter-for-Global-Poverty.pdf

Lall, S. V., & Wahba, S. (2021). Crowded cities: New methodology in COVID-19 risk assessment. *Sustainability, 13*, 7167. https://doi.org/10.3390/su13137167

Malin, M. (2021). *Increasing government effectiveness using the SDGs*. City Playbooks for Advancing the SDGs. Brookings Center for Sustainable Development. https://www.brookings.edu/wp-content/uploads/2021/07/City-playbook_Helsinki.pdf

Martinelli, F., & Lindner, A. (2021). How far how far are the metropolitan areas in Brazil from achieving the sustainable development goals? An analysis based on SDG dashboards. *SN Social Sciences, 1*(103). https://doi.org/10.1007/s43545-021-00131-8

Masuda, H., Okitasari, M., Katramiz, T., Shimizu, H., Kawakubo, S., & Kataoka, Y. (2021). SDGs mainstreaming at the local level: Case studies from Japan. *Sustainability Science, 16*, 1539–1562. https://doi.org/10.1007/s11625-021-00977-0

Mohieldin, M., & Shehata, M. (2021). *The SDGs as an operational framework for post COVID-19 response and recovery*. AD-inister N°. 38 enero - junio 2021 pp. 5–42 · ISSN 1692-0279 · eISSN 2256-4322.

National League of Cities. (2020). *What COVID-19 means for city finances*. Final Report. https://covid19.nlc.org/wp-content/uploads/2020/06/What-Covid-19-Means-For-City-Finances_Report-Final.pdf

Negreiros, P., Furio, V., Falconer, A., Richmond, M., Yang, K., Jungman, L., Tonkonogy, B., Novikova, A., Pearson, M., & Skinner, I. (2021). *The state of the cities climate finance. Part 1: The landscape of urban climate finance*. Climate Policy Initiative. https://www.citiesclimatefinance.org/wp-content/uploads/2021/06/Part-1-l-The-Landscape-of-Urban-Climate-Finance-FINAL.pdf

Nieuwenhuijsen, M. (2020). *Urban and transport planning pathways to carbon neutral, livable and healthy cities. A review of the current evidence*. https://doi.org/10.1016/j.envint.2020.105661

OECD. (2020). *A territorial approach to the sustainable development goals*. https://www.oecd-ilibrary.org/sites/e86fa715-en/1/1/1/index.html?itemId=/content/publication/e86fa715-en&_csp_=6d42c3e04d6f0b50b2439e764fde242b&itemIGO=oecd&itemContentType=book

Ortiz-Moya, F., Koke, H., Ota, J., Kataoka, Y., & Fujino, J. (2020). *State of the voluntary local reviews 2020: Local action for global Impact in achieving the SDGs*. Institute for Global Environmental Strategies (IGES): Kanagawa, Japan. ISBN: 978-4-88788-243-0.

Pagiola, S., & Rios, A. R. (2013). *Evaluation of the impact of payments for environmental services on land use change in Quindío, Colombia*. World Bank. https://openknowledge.worldbank.org/handle/10986/21122. License: CC BY 3.0 IGO.

Reid, C. (2020, April 22). *Paris to create 650 kilometers of post-lockdown cycleways*. https://www.forbes.com/sites/carltonreid/2020/04/22/paris-to-create-650-kilometers-of-pop-up-corona-cycleways-for-post-lockdown-travel/?sh=3a828354d404

Ribeiro Lucinda, C. (2017). *Economies of scale and scope in the sanitation sector*. https://doi.org/10.12660/bre.v37n22017.47626

Rubinyi, S., Goldblatt, R., & Park, H. (2020). *Nighttime lights are revolutionizing the way we understand COVID-19 and our world*. The World Bank. https://blogs.worldbank.org/sustainablecities/nighttime-lights-are-revolutionizing-way-we-understand-covid-19-and-our-world

The Economist. (2020, June 18). A mayor is reforming Sierra Leone's rotten property tax. *The Economist*. https://www.economist.com/middle-east-and-africa/2020/06/18/a-mayor-is-reforming-sierra-leones-rotten-property-tax

Tholley, D. A. (2021). *Empowering young people to contribute to Freetown's green recovery*. Urbanet. https://www.urbanet.info/empowering-young-people-freetown-green-recovery/

Tobin, S., & Zaman, A. (2022). Regional cooperation in waste management: Examining Australia's experience with inter-municipal cooperative partnerships. *Sustainability, 14*(3), 1578. https://doi.org/10.3390/su14031578

UCLG. (2019). *Towards the localization of the SDGs. Local and Regional Governments' Report Towards the 2019 HLPF*. 3rd Report. Global Task Force of Local and Regional Governments. https://www.uclg.org/sites/default/files/towards_the_localization_of_the_sdgs_0.pdf

UCLG. (2020). *The sustainable development goals*. What Local Governments Need to Know. https://www.uclg.org/sites/default/files/the_sdgs_what_localgov_need_to_know_0.pdf

UGLG & UN-Habitat. (2020). *Guidelines for voluntary local reviews. A comparative Analysis of Exinting VLRs*. https://www.uclg.org/sites/default/files/uclg_vlrlab_guidelines_2020_volume_i.pdf

United Nations. (2015). *Transforming our world: The 2030 agenda for sustainable development*. Resolution adopted by the General Assembly on 25 September 2015. A/RES/70/1. U.N. General Assembly, Seventieth Session. Agenda items 15 and 116. United Nations.

United Nations. (2019). *The sustainable development goals report 2019*. https://unstats.un.org/sdgs/report/2019/The-Sustainable-Development-Goals-Report-2019.pdf

United Nations. (2021). *The sustainable development goals report 2021*. https://unstats.un.org/sdgs/report/2021/The-Sustainable-Development-Goals-Report-2021.pdf

United Nations. (2022). *Payments for environmental services program: Costa Rica*. https://unfccc.int/climate-action/momentum-for-change/financing-for-climate-friendly-investment/payments-for-environmental-services-program

USAID. (2021). *Administrator Samantha Power on a new vision for global development*. https://www.usaid.gov/news-information/speeches/nov-4-2021-administrator-samantha-power-new-vision-global-development

Valencia, S. C., Simon, D., Croese, S., Nordqvist, J., Oloko, M., Sharma, T., Taylor Buck, N., & Varsace, I. (2019). Adapting the sustainable development goals and the new urban agenda to the city level: Initial reflections from a comparative research project. *International Journal of Urban Sustainable Development, 11*(1), 4–23. https://doi.org/10.1080/19463138.2019.1573172

Wahid, A. N. M. (1994). The Grameen Bank and poverty alleviation in Bangladesh: Theory, evidence and limitations theory, evidence and limitations. *American Journal of Economics and Sociology, 53*(1), 1–15. https://doi.org/10.1111/j.1536-7150.1994.tb02666.x

White, H., Menon, R., & Waddington, H. (2018). *Community-driven development: Does it build social cohesion or infrastructure? A mixed-method evidence synthesis*. International Initiative for Impact Evaluation. https://www.3ieimpact.org/sites/default/files/2019-01/wp30-cdd_0.pdf

Whittle, N. (2020, July 17). Welcome to the 15-minute city. *The Financial Times*. https://www.ft.com/content/c1a53744-90d5-4560-9e3f-17ce06aba69a

World Bank. (2022). *The impact of COVID-19 on municipal finance*.

World Bank & International Monetary Fund. (2013). *Global monitoring report 2013: Rural-urban dynamics and the millennium development goals*. World Bank. https://openknowledge.worldbank.org/handle/10986/13330

World Bank. (2017). *Tunisia WASH and poverty diagnostic*. Phase 2 Report: WASH accounts. https://documents1.worldbank.org/curated/en/760031499338439264/pdf/ACS17905-V2-WP-P153680-PUBLIC.pdf

World Bank. (2019). *Kenya's devolution*. https://www.worldbank.org/en/country/kenya/brief/kenyas-devolution

World Bank. (2020). *Poverty and shared prosperity 2020: Reversals of fortune*. World Bank. https://doi.org/10.1596/978-1-4648-1602-4.License:CreativeCommonsAttributionCCBY3.0IGO

Zinkernagel, R., Evans, J. & Lena Neij. (2018), Applying the SDGs to Cities: Business as usual or new dawn? *Sustainability, 2018*(10), 3201. https://doi.org/10.3390/su10093201

CHAPTER 5

Never Let a Good Crisis Go to Waste

Abstract COVID-19 not only exposed complicated challenges for society, markets, the planet, and businesses, but it also hastened several trends, such as digital transformation, additive manufacturing, the pursuit of social and environmental sustainability, and climate change risk awareness. The world is currently at a tipping point in history, and there is an urgent need to hasten the implementation of the Sustainable Development Goals (SDGs). Already half of the 2015–2030 Agenda has been completed, and the deadline is soon approaching.

Crises can lead to transformation by removing long-term barriers. Furthermore, the adaptive phase of crisis leadership addresses root problems and fosters resilience. Responding well to threats such as COVID-19, and more recently the war in Ukraine, could lead to a shift in thinking about how to fix systemic flaws and key sources of risk. This will in turn allow for better long-term policy implementation and major structural changes.

Keyword SDG; partnerships; alliances; climate action; Build Forward Better; social contract; acceleration; multi-actor; governance

M. Mohieldin et al., *Business, Government and the SDGs*,
https://doi.org/10.1007/978-3-031-11196-9_5

A Turning Point in History

The world is at a historical crossroads. The pandemic set back much progress on poverty alleviation. Indeed, it is estimated to have pushed an additional 97 million people in extreme poverty in 2020, an increase of 11.6% year-on-year after years of steady decline (Gerszon Mahler et al., 2021). This setback undoes the equivalent of five years of progress. Time is running out to take on the type of policies and commitments necessary to avert a climate crisis. As a result, the ability to combine policies and interventions to achieve both recovery from COVID-19 and transformative climate action is critical to the future of humankind.

Undoubtedly, there is rising awareness of the critical importance of sustainable development, and the SDGs are being incorporated into the lexicon of government, business, spirituality, science, and community leaders. Despite all this, there has currently been limited progress toward the SDGs and long-term sustainability (Allen et al., 2021; Temmer & Jungcurt, 2021). In fact, the SGDs were off-track even before the pandemic began (Guterres, 2021; Guterres, 2022). Now, the world is nearly at the midway point between the launch date in 2015 and the target of 2030. To elaborate, Figure 8 illustrates how certain SDG goals are making good progress while others show little improvement and even deterioration (Fig. 5.1).

After more than two years since the emergence of COVID-19, which upended many elements of day-to-day life and negatively impacted people's livelihoods, we as a society are coming to terms with learning how to live with the pandemic (Lavine et al., 2022). As of now, while new variants of the virus continue to evolve and much work is still to be done concerning global vaccination access, it is important to find adaptive solutions to overcome the challenges of achieving the SDGs.

COVID-19 compelled the world to re-examine the traditional roles and spheres of influence of government, the business sector, and communities, and their relationships. The 2030 Agenda of SDGs is the best way to look at these roles and relationships, and assess whether they are financially, environmentally, and socially sustainable.

Perhaps the most challenging issue is that the SDGs are not garnering the level of attention needed to catalyze the scale of action needed to meet the 2030 targets (Tonne, 2021). In fact, the urgency of climate action appears to have displaced the attainment of the SDGs as the main imperative of our generation, which constitutes a critical driver of development

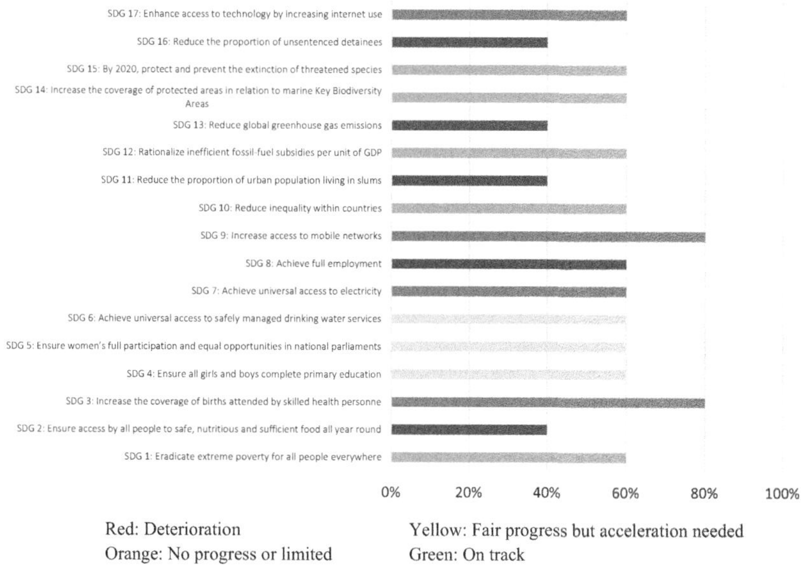

Fig. 5.1 Progress toward the SDGs on selected targets (*Source* Authors based on UNSTAST [2021])

action. For instance, a recently published article in Foreign Policy argued that rich donor countries, as the key shareholders of the World Bank and the International Monetary Fund (IMF), are increasingly shifting the two institutions' priorities to reducing carbon emissions at the expense of their traditional focus on poverty reduction and economic growth in the global south (Ramachandran & Baker, 2022).

Climate Action and Development as Mutually Unexclusive

Climate action and development, however, are not intended to be at odds (Hallegate, 2022). Evidence increasingly suggests that climate action need not come at the expense of economic growth (Hallegate, 2022). The right choice of climate policies, prioritizing of nature, and drawing of existing technologies, may boost productivity and efficiency for low-income countries (Hallegate, 2022). However, to avoid a disconnection between climate action and economic growth and development, an integrated approach is needed, which is found within the framework of the SDGs.

It is clear that with the situation as it currently stands, national governments alone cannot attain the SDGs or take the transformative climate action needed to maintain emissions within a 1.5° C in line with the Paris Agreement (UNFCCC, 2015). The Time Magazine's cover for April 2022, for one, noted how governments are failing and businesses are taking on the climate fight (Wordland, 2022). If the world stands a chance at achieving the SDGs by 2030, collective action is needed involving all levels of government, the private sector, and civil society (Gonzalez-Perez, 2022; Horan, 2022; Monkelbaan, 2019). Despite the fact that national governments are the main actors responsible for the achievement of the SDGs, the 2030 Agenda emphasizes the importance of having them act in a manner that creates incentives and enabling environments. This is necessary in order to foster both whole-of-government and multistakeholder approaches, with a focus on attracting the private sector.

The World Bank's newly introduced diagnostic tool—Country Climate and Development Report (CCDR)—was designed to emphasize that climate action needs to be integrated within development pathways for any given country in order to be effective, which comes as an extra emphasis that climate change and development, as reflected in the SDGs, are interconnected. Both agendas also require sufficient scaling up of adequate finance to meet their ambitions. Given the interlinkages between the two and the thus-far quite limited financial flows directed toward them, increasing the effectiveness of climate and development finance necessitates instilling coherence between these two strands of finance in a matter that maximizes synergies and minimizes trade-offs (OECD, 2019). Additionally, it is important to deepen the financial effectiveness and ownership of these two types of finance by increasing alignment between the donor countries' finance and the priorities of recipient countries (Iacobuţă et al., 2022). Both climate and development finance are sourced in taxpayers and work to finance the given climate or developmental-related interventions in recipient states. Thus, to ensure that this aligns with the priorities of the recipients, both strands of climate and development finance need to be communicated through national budgets.

Furthermore, private finance and investment are also required to leverage both climate and development finance; a theme that was stressed in COP26 and is among the top priorities to be planned on the COP27 agenda (Mohieldin, July 2022). However, mobilizing private finance requires a strong signaling system supported by policies and regulatory frameworks that promote the attraction of private financial flows to the highest priority sectors. This, in turn, needs to be implemented alongside a comprehensive framework that coordinates the different financial flows available. The INFF provides this alignment framework through working to build a financing strategy that mobilizes funding for key priorities (UN DESA, 2022).

The Importance of Collaboration and Coordination

To complement the national government and the private sector, local stakeholders such as local governments and CBOs are needed for last-mile delivery (Gonzalez-Perez, 2022; Rashed & Shah, 2021). Subnational governments in particular should take the lead in the localization of the SDGs, including defining interventions and targets, as well as implementing and monitoring them at the local level (Guha & Chakrabarti, 2019; Reddy, 2016). At the same time, NGOs and CBOs could play an important role in accessing populations that are hard to reach, population, so as to "leave no one behind."

In addition to the all-hands-on-deck approach, partnerships between public and private stakeholders are especially needed to harness the efforts of the private sector in unchartered areas. Public–private partnerships (PPP), guarantees, or other forms of credit enhancement are all relevant instruments to crowd in private investment in infrastructure and service delivery by reducing the level of perceived risk.

In addition to collaboration and partnerships, coordination is equally important to ensure that actions needed by one entity that have a catalytic effect on other stakeholders can be implemented early on to unlock multiple interventions. A particularly relevant example is the role of national governments in setting up an enabling environment for private sector and local governments to act (e.g., regulations governing public–private partnerships or contract enforcement and decentralization policies and revenue mobilization legislation, respectively).

On a positive note, there is strong emerging leadership from businesses and local governments on issues that were recently not part of their traditional remit, such as matters concerning global public goods and social issues. Indeed, both businesses and local governments are assuming greater leadership in issues of sustainability, climate action, and refugees, oftentimes compensating for inadequate or insufficient national efforts in these domains. An example would be the response to US national government's withdrawal from the Paris Agreement, when 468 US Mayors committed to achieving the Agreement's targets within their local jurisdiction (Climate Mayors, 2019). Similarly, cities around the world including in Lebanon, Jordan, Turkey, Germany, and Poland welcomed large numbers of refugees fleeing wars in Syria and Ukraine, and offered sanctuary even at the great cost of overwhelming their state's service delivery and housing systems.

However, much of the global efforts today seem to be more focused on climate change than on the other SDGs, which reflect broader sustainability challenges, of which climate change is only one of 17 objectives. Although, there is no denying that climate change is the most critical issue as if current emissions trends continue unabated, progress on all other SDGs,—especially poverty and inequality—will be significantly derailed. That said, there is more to sustainability than climate change. The SDGs' own framework emphasizes social inclusion and "leaving no one behind" more so than on a just transition, which is a part of climate change action plans.

The Necessity of Immediate Collaborative Action in Order to Build Forward Better

To be able to accelerate the achievement of the SDGs and get to the scale of action needed to meet the targets by 2030, much still needs to be done. Urgency is particularly needed concerning the last-mile delivery of the SDGs—including infrastructure, services, and jobs, among other priorities—which will require the active involvement of both businesses and local governments to complement national governments' efforts. A key prerequisite for this is for national governments to put in place an enabling regulatory environment that facilitates and incentivizes the actions of local governments and businesses.

For local governments to act, they must be empowered to take needed actions and have access to adequate resources for implementation. This in turn inevitably requires meaningful decentralization, both administrative/political and fiscal in nature. Additionally, local governments will need incentives and financial support to influence and complement their resource allocation priorities. This is alongside the need for capacity building to strengthen their planning, service delivery, monitoring and reporting abilities. Enabling a more rational, efficient, and cost-effective process of service delivery would benefit from an alignment of local governments' jurisdiction with the functional areas of service delivery, given the common disconnection between municipalities' administrative boundaries and functional service delivery in the context of metropolitan areas. Lastly, strengthening local governments' resilience against disasters, pandemics, and other shocks will be essential to avoid more setbacks in achieving the SDGs.

Businesses also have a critical role to play in the achievement of the SDGs. In certain aspects and geographies, a shifting consumer demand to emphasize the importance of sustainability will be what is needed to trigger action (e.g., more willing to pay for fair trade goods). In many other areas, however, government interventions will be needed to incite a commensurate business response. This includes ensuring the necessary protections to intellectual property together with public subsidies and incentives for companies to invest in new technologies and solutions to SDG challenges (e.g., enhancing access to affordable renewable energy, COVID-19 vaccines, or setting up public–private partnerships for service delivery). Equally important is the role of governments in setting up the regulatory framework regarding pricing and enforcement to tackle negative social and environmental externalities (e.g., carbon pricing, pollution tax, etc.).

The private and public sectors in particular must change how they work together to achieve a transformative impact. Public–private partnerships should be about 'co-investment, sharing risks and sharing rewards' (Mazzucato, 2021, p. 129). The private sector's contribution would not be 'corporate social responsibility or charity, but through the value chain,—where money is made while supporting society aims—and through investment with public purpose' (Mazzucato, 2021).

Even with the involvement of local governments and businesses, there will still be those excluded, including disabled people, indigenous communities, refugees, as well as religious, ethnic, and sexual minorities, among other vulnerable and marginalized groups. Those facing intersectional exclusion will be even more difficult to reach. As such, if the SDGs' promise of "leaving no one behind" is to be achieved, there needs to be innovative solutions, including the involvement of multiple non-traditional stakeholders such as NGOs, civil society, social enterprises, and

CBOs. Involving communities proactively in the prioritization, design, and implementation of solutions to the SDGs—as opposed to just acting as beneficiaries—is also important in ensuring that solutions effectively respond to the relevant needs and demands. It also works to confirm that resulting outcomes are sustainable and integrate community involvement in both the operation and maintenance.

The Critical Leadership of National Governments for MultiStakeholder Delivery

The main role in achieving the SDGs will continue to rest with national governments, whose leadership is critical in implementing the key elements for multistakeholder delivery. This includes: (a) enabling regulatory environments; (b) incentives for action; (c) reforms to facilitate private sector, local government, and civil society involvement and factor in negative externalities; (d) structures to ensure effective coordination among stakeholders and levels of government; and (e) putting in place the support that is needed for the hardest to reach population.

There is no doubt that achieving the SDGs will be a very costly endeavor, especially considering the need to offset the negative impacts of COVID-19. A World Bank study estimates that low- and middle-income countries will require an annual investment of USD 1.5–2.7 trillion during 2015–2030 to meet infrastructure-related SDGs, an amount that is equivalent to 2.5–8.2% of their respective GDPs (Rozenberg & Fay, 2019). Preliminary estimates show that the pandemic has exacerbated the SDG financing gap by up to USD 4.2 trillion (70%) in developing countries, as caused by the joint effect of the additional needs for COVID-19 spending and the drop in external private sources (OECD, 2020). Appropriate budgeting by national governments is therefore a critical first step to achieving the goals (Mohieldin & Shehata, 2021).

Elements of the 2030 Agenda pertaining to poverty, health, education, security, and economic development, are already incorporated in the national policies, plans, budgets, and accountability systems of most governments. However, the SDGs represent a more comprehensive and structured framework for these aspects that is further supported by measurable and follow-up dimensions. At the heart of the SDGs is integration and interactions between the goals that call for coherence and coordination in national governance mechanisms. However, solely becoming a part of a country's national policy framework is insufficient for progressing toward the SDGs. The goals should become integrated into national budgetary frameworks to ensure that policy priorities are embedded in the budget, which represents the 'strongest domestic legal basis that reflects countries' commitments to 2030 Agenda and its related SDGs' (UNDP, 2020). Otherwise, it is very possible for national development agendas to make commitments to the SDGs while public budgets remain unchanged. As such, budgets are powerful instruments to ensure that SDG plans are implemented and this yields its intended outcomes. This particularly applies when targets are measurably represented through concrete budgetary allocations and reports, as well as when budget execution evaluations feed into the decision-making process.

The COVID-19 pandemic highlighted two key issues. First, shocks and stresses such as pandemics, disasters from natural hazards, or conflicts can occur at any point in time, and they cause serious setbacks in development. As a result, it is critical to invest in strengthening resilience to avoid derailing progress toward the achievement of the SDGs. Secondly, COVID-19 exposed the extent of existing inequalities including access to housing, health, education, safety nets, infrastructure, municipal and social services, and the precarious living situations of many, such as informal sector workers or slum dwellers, among others.

Failure to tackle these inequalities will leave many people behind, and a failure to achieve the SDGs. Moreover, there will constantly be a threat of pandemic, as viruses continue to mutate spread, even with the availability of vaccines. Similarly, a failure to expand access to economic development opportunities, to prioritize peace and stability, and to invest in climate change adaption in the most deprived global regions will deplete natural resources and negatively impact livelihoods. This will result in further setbacks to the achievement of the SDGs, as well as exacerbating the flows of conflict- and climate-induced migration, with a continued tragic loss of lives.

One of the silver linings of the pandemic was to highlight the importance and transformative potential of digitalization, which may accelerate the necessary infrastructure deployment for the SDGs. Increased adoption of digitalization together with disruptive technologies can be advantageous in the achievement of the goals through potentially reducing material waste, mitigating carbon emissions, and accelerating access to education among other possibilities. However, this requires ensuring equitable and broad access to digital infrastructure in order to avoid the risk of a digital divide that leaves behind vulnerable groups such as the poor, elderly, and those with limited education.

Accountability Systems for the 2030 Agenda

Despite the fact that the SDGs do not have a binding global accountability system, different actors are responsible for their contributions to the 2030 Agenda and its economic, social, and environmental dimensions through their accountability systems when engaging with relevant stakeholders. To elaborate, while national governments have committed to the achievement of the 2030 Agenda, it was emphasized that governments should adapt these global goals to their national contexts, taking into account their own countries' sets of constraints and opportunities. In doing so states have become accountable to their populace and other stakeholders for which these commitments have been made. Governments are thus responsible for putting in place the needed mechanisms, policies, and resources to realize these now-national agendas and plans. This adaptation of global commitments into national targets entails ownership and answerability (UN DESA, 2015). Accordingly, national governments are responsible for these global-turned-national commitments via their respective accountability mechanisms, which tend to differ according to a given country's political and governance systems. However, and as repeatedly emphasized throughout this text, the most crucial mechanism in holding the government accountable is the national budget (UNDP, 2020). Hence, having the national budget aligned with the SDGs and reflected in a country's national plans and strategies facilitates an accountability process that is informed by state priorities and allows for increased evidence-based accountability.

Among the main aspects emphasized in the 2030 Agenda is the importance of last-mile delivery at the local level and the acknowledgment of territorial specificities in the development process. Local and regional governments (LRGs) provide the basic frontline services required for development and represent the level of government closest to the people. As such, they act as a touchstone for SDG accountability as reflected in a country's development strategy. In that sense, the localization approach to development has gained momentum as a process that maintains the benefits of centralization through aligning national development strategies with the needs of the local communities, as well as that which promotes accountability and transparency at the local level. The latter being the result of decision-making that occurs at a level close to target populations and is motivated by a solid understanding of the local context and demands of the local communities involved (Elmassah & Mohieldin, 2020).

ESG considerations capture the manner in which the business sector interacts with and impacts its ecosystem. Given the scope of this sector, the ESG mechanism represents the most adequate means of holding businesses accountable for their actions (Gapper, 3 June 2022). ESG issues have thus been advanced in many ways to enhance their transparency and accountability roles. New legislations have been introduced to enhance ESG disclosure requirements, and increased legal actions are being taken regarding ESG-related matters as a backlash to companies greenwashing and making false or exaggerated ESG accomplishment claims (Temple-West, 23 May 2022). Companies are also under huge pressure from investors, customers, and employees to deliver their commitments to sustainability and diversity (PwC, 2021). Additionally, by their composition and legal mandate, companies are accountable to their general assemblies and shareholders. Beyond the ESG considerations, some businesses have gone the extra mile of committing to the 2030 Agenda by forming businesses alliances for the SDGs and climate action. For example, the Glasgow Financial Alliance for Net Zero (GFANZ) brings together more than 450 member firms from across the global financial sector, representing more than $130 trillion in assets under management and advice which are committed to broadening, deepening, and raising net-zero ambitions across the financial system. Similarly, the Global Investors for Sustainable Development (GISD) Alliance consists of 30 leaders of major financial institutions and corporations who seek to deliver concrete solutions that scale-up long-term finance and investment in sustainable development. Furthermore, the signatories of the Principles for Responsible Investment (PRI) are also leading institutional investors for responsible investment. Businesses and investors that become members of these kinds of alliances and networks are obliged to hold themselves globally accountable for the advancement of the 2030 Agenda.

Updating the Global Social Contract

In conclusion, the world needs a new global social contract if it is to achieve the SDGs by 2030 despite the two major crises of COVID-19 and climate change. There is a current urgency to act and ensure that the recovery from the pandemic is aligned with the necessary climate action in order to achieve a green, resilient, and inclusive future. After all, there is no Planet B.

Acknowledgements Authors recognize Haajar Abu Ismail's contribution in editing this chapter.

References

Allen, C., Metternicht, G., & Wiedmann, T. (2021). Priorities for science to support national implementation of the sustainable development goals: A review of progress and gaps. *Sustainable Development, 29*(4), 635–652.

Climate Mayors (2019, November 27). 468 US Climate Mayors commit to adopt, honor and uphold Paris Climate agreement goals. Available at: https://climatemayors.org/actions-paris-climate-agreement/

Dalby, S., Horton, S., & Mahon, R. (2019). Global governance challenges in achieving the Sustainable Development Goals: Introduction. In *Achieving the sustainable development goals* , (pp. 1–18). Routledge.

Elmassah, S., & Mohieldin, M. (2020). Digital transformation and localizing the sustainable development goals (SDGs), *Ecological Economics, 169*, 106490. Elsevier. https://doi.org/10.1016/j.ecolecon.2019.106490

Gapper, J. (2022, June 3). *Greenwashing is tempting for CEOs who tell stories. Financial Times* [online]. Available from: https://www.ft.com/content/40192e1b-378d-4eaa-97f2-39491d276faf

Gerszon Mahler, Daniel, et al. (2021). Updated estimates of the impact of COVID-19 on global poverty: Turning the corner on the pandemic in 2021? Available at: https://blogs.worldbank.org/opendata/updated-estimates-impact-covid-19-global-poverty-turning-corner-pandemic-2021 (Last accessed August 11, 2021)

Gonzalez-Perez, Maria Alejandra (2022) *Regenerative and sustainable futures for Latin America and the Caribbean: Collective action for a region with a better tomorrow*. Emerald: Bingley, UK

Guha, J., & Chakrabarti, B. (2019). Achieving the Sustainable Development Goals (SDGs) through decentralization and the role of local governments: A systematic review. *Commonwealth Journal of Local Governance, 22*, 1–21.

Guterres, A. (2021). The sustainable development goals report 2021. UN Stats. https://unstats.un.org/sdgs/report/2021/The-Sustainable-Development-Goals-Report-2021.pdf

Guterres, A. (2022, February 10). Not a moment to lose, in race to meet sustainable development goals. UN News; news.un.org. https://news.un.org/en/story/2022/02/1111672

Hallegatte, S. (2022, April 13). Getting it right on development: We do not have to choose between people and climate. World Bank Blogs. Available at: https://blogs.worldbank.org/climatechange/getting-it-right-development-we-do-not-have-choose-between-people-and-climate

Horan, D. (2022). A framework to harness effective partnerships for the sustainable development goals. *Sustainability Science*. https://doi.org/10.1007/s11625-021-01070-2

Iacobuţă, G.I., Brandi, C., Dzebo, A. & Duron, S. D. E. (2022). Aligning climate and sustainable development finance through an SDG lens. The role of development assistance in implementing the Paris Agreement. *Global Environmental Change*, *74*, 102509. https://doi.org/10.1016/j.gloenvcha.2022.102509.

Lavine, J. S., Bjornstad, O. N., & Antia, R. (2022). Immunological characteristics govern the transition of covid-19 to endemicity. *Science*, *371*, 741–745. Scientific Reports, 12, 4731.

Mazzucato, M. (2021). *Mission economy: A moonshot guide to changing capitalism*. Harper Business.

Mohieldin, M (2022, July). *Vicious circles of climate change*. Ahram Online. [online]. Available from: https://english.ahram.org.eg/NewsContentP/50/470796/AlAhram-Weekly/Vicious-circles-of-climate-change.aspx

Mohieldin, M., & Shehata, M. (2021). The SDGs as an operational framework for post COVID-19 response and recovery. *AD-Minister*. *38*, 5–42 https://doi.org/10.17230/Ad-minister.38.1

Monkelbaan, J. (2019). *Governance for the sustainable development goals: Exploring an integrative framework of theories, tools, and competencies*. Springer.

OECD. (2020). Global outlook on financing for sustainable development 2021: A new way to invest for people and planet. *OECD Publishing, Paris*,. https://doi.org/10.1787/e3c30a9a-en

OECD (2019). *Recommendation of the council on policy coherence for sustainable development*. OECD/LEGAL/0381'. Organization for Economic Co-operation and Development, Paris

PwC (2021). 2021 ESG Consumer intelligence series: *Beyond compliance: Consumers and employees want business to do more on ESG—How business can close the expectations gap*. Available from: https://www.pwc.com/us/en/services/consulting/library/consumer-intelligence-series/consumer-and-employee-esg-expectations.html

Ramachandran, Vijaya and Arthur Baker (2022, April 11). The World Bank and IMF are getting it wrong on climate change. Foreign policy. Available at: https://foreignpolicy.com/2022/04/11/the-world-bank-and-imf-are-getting-it-wrong-on-climate-change/

Rashed, A. H., & Shah, A. (2021). The role of private sector in the implementation of sustainable development goals. *Environment, Development and Sustainability, 23*(3), 2931–2948.

Reddy, P. S. (2016). Localising the sustainable development goals (SDGs): the role of local government in context. Available at: https://repository.up.ac.za/bitstream/handle/2263/58190/Reddy_Localising_2016.pdf?sequence=1

Rozenberg, J., & Fay, M. (Eds.). (2019). *Beyond the gap: How countries can afford the infrastructure they need while protecting the planet*. World Bank Publications.

Temmer, J., & Jungcurt, S. (2021). Data for good for all: Enabling all communities to track progress toward SDG implementation. In *Promoting the sustainable development goals in North American cities* (pp. 97–108). Springer, Cham.

Temple-West, P. (2022, 23 May). *Watchdogs tackle the murky world of greenwash. Financial Times* [online]. Available from: https://www.ft.com/content/85ca91c0-036f-47a1-9a76-3da48caf6851

Tonne, C. (2021). Lessons from the COVID-19 pandemic for accelerating sustainable development. *Environmental Research, 193*, 110482.

United Nations Department of Economic and Social Affairs (UN DESA) (2022). *Integrated National Financing Frameworks—moving towards financing policy integration*. UN DESA Policy Brief No. 124. Available from: https://www.un.org/development/desa/dpad/publication/un-desa-policy-brief-124-integrated-national-financing-frameworks-moving-towards-financing-policy-integration/

United Nations Department of Economic and Social Affairs (UN DESA) (2015). *Transitioning from the MDGs to the SDGs: Accountability for the Post-2015 Era*. CDP Background Paper No. 25. Available from: https://www.un.org/development/desa/dpad/wp-content/uploads/sites/45/publication/CDP-bp-2015-25.pdf

United Nations Development Programme (UNDP) (2020). *Budgeting for the sustainable development goals: Aligning domestic budgets with the SDGs: A Guidebook*. https://sdgfinance.undp.org/sites/default/files/UNDP%20Budgeting%20for%20the%20SDGs%20-%20Guidebook_Nov%202020.pdf

UNFCCC (2015). The Paris Agreement. Available online at: https://unfccc.int/sites/default/files/english_paris_agreement.pdf

UNSTATS (2021). Sustainable development goals progress chart. Available at: https://unstats.un.org/sdgs/report/2021/progress-chart-2021.pdf

Wordland, J. (2022) Governments are failing. How business is taking on the climate fight. *Time*. https://time.com/6166178/earths-future-big-business/?utm_source=email&utm_medium=email&utm_campaign=email-share-article&utm-term=climate_climate

Index

M. Mohieldin et al., *Business, Government and the SDGs*,
https://doi.org/10.1007/978-3-031-11196-9